High-Frequency Trading and Probability Theory

East China Normal University Scientific Reports

ISSN: 2382-5715

Chief Editor
Weian Zheng
Changjiang Chair Professor
School of Finance and Statistics
East China Normal University, China
Email: financialmaths@gmail.com

Associate Chief Editor
Shanping Wang
Senior Editor
Journal of East China Normal University (Natural Sciences), China
Email: spwang@library.ecnu.edu.cn

Vol. 1   High-Frequency Trading and Probability Theory
*by Zhaodong Wang (East China Normal University)
and Weian Zheng (East China Normal University)*

Vol. 2   School Mathematics Textbooks in China:
Comparative Study and Beyond
by Jianpan Wang (East China Normal University)

East China Normal University Scientific Reports–Vol. 1

High-Frequency Trading and Probability Theory

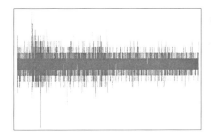

Zhaodong Wang
Weian Zheng

East China Normal University, China

World Scientific

NEW JERSEY · LONDON · SINGAPORE · BEIJING · SHANGHAI · HONG KONG · TAIPEI · CHENNAI

Published by

World Scientific Publishing Co. Pte. Ltd.
5 Toh Tuck Link, Singapore 596224
USA office: 27 Warren Street, Suite 401-402, Hackensack, NJ 07601
UK office: 57 Shelton Street, Covent Garden, London WC2H 9HE

Library of Congress Cataloging-in-Publication Data
Wang, Zhaodong.
  High-frequency trading and probability theory / Zhaodong Wang, East China Normal University, China and Weian Zheng, East China Normal University, China.
   pages cm. -- (East China Normal University scientific reports ; v. 1)
  Includes bibliographical references and index.
   ISBN 978-9814616508 (hardcover : alk. paper) -- ISBN 978-9814616515 (pbk. : alk. paper)
  1. Investment analysis. 2. Portfolio management. 3. Electronic trading of securities. I. Title.
  HG4529.W36 2014
  332.64'20285--dc23
                               2014024086

British Library Cataloguing-in-Publication Data
A catalogue record for this book is available from the British Library.

Copyright © 2015 by World Scientific Publishing Co. Pte. Ltd.

All rights reserved. This book, or parts thereof, may not be reproduced in any form or by any means, electronic or mechanical, including photocopying, recording or any information storage and retrieval system now known or to be invented, without written permission from the publisher.

For photocopying of material in this volume, please pay a copying fee through the Copyright Clearance Center, Inc., 222 Rosewood Drive, Danvers, MA 01923, USA. In this case permission to photocopy is not required from the publisher.

In-house Editors: Sandhya Venkatesh/Chitralekha Elumalai.

Typeset by Stallion Press
Email: enquiries@stallionpress.com

Printed in Singapore

Contents

Foreword

Preface

About the Authors

1. Introduction
2. Market Microstructure
    - 2.1 Trading Products . . . .
    - 2.2 Trading Model . . . .
        - 2.2.1 Continuous Trading . . . .
        - 2.2.2 Auction . . . .
    - 2.3 Market Data Information . .
    - 2.4 Trading Interface . . . . . .
    - 2.5 Risk Control . . . .
    - 2.6 Transaction Costs . .
    - 2.7 Differences with West
3. Some Basic HFT Strategi
    - 3.1 General . . . .
    - 3.2 Arbitrage . .
        - 3.2.1 Def
        - 3.2.2 Γ
        - 3.2.3
    - 3.3 Ticker Tap
    - 3.4 Market Making

	3.5	Event Driven		54
	3.6	Other Basic Strategies		56
4.	IT System			59
	4.1	Challenges		59
	4.2	Trading System Design		61
		4.2.1	Trading Interface	62
		4.2.2	Trading Process Control	66
		4.2.3	Risk Control and Surveillance	70
		4.2.4	Strategy Implementation	75
		4.2.5	Monitoring	77
	4.3	Environment		77
		4.3.1	Programming Language	77
		4.3.2	Server and Operation System Selection	78
		4.3.3	Network Environment	79
	4.4	Core Technologies		81
		4.4.1	Single-Process vs. Multi-Process	81
		4.4.2	Code Optimization	82
		4.4.3	Memory Management	83
		4.4.4	Managing CPU Cache	83
5.	Stationary Process and Ergodicity			85
	5.1	Some Basics of Probability Theory		85
		5.1.1	Probability Space	86
		5.1.2	Random Variables	88
		5.1.3	Conditional Probability	89
		5.1.4	Two Main Theorems	90
			5.1.4.1 The strong law of large number	91
			5.1.4.2 The central limit theorem	92
	5.2	Stochastic Process		92
		5.2.1	Examples of Stochastic Processes	92
		5.2.2	Stationary Process and Ergodic Theory	93
			5.2.2.1 The strong ergodic theorem	95
		5.2.3	Testing Stationarity	97
		5.2.4	Semi-Martingales and Filtering Problem	98
		5.2.5	Stationary Process as Noises	100

	5.3	Time Series Analysis	101
	5.4	Pair-Trading Revisited	104
6.	Stationarity and Technical Analysis		109
	6.1	Technical Analysis	111
	6.2	Logarithmic Return is Stationary	113
	6.3	Moving Average and Exponential Moving Average	115
	6.4	Bollinger Bands	120
	6.5	Moving Average Convergence–Divergence	123
	6.6	Rate of Change	126
	6.7	Relative Strength Index	127
	6.8	Stochastic Oscillators	129
	6.9	Directional Movement Index	133
	6.10	Parabolic SAR	138
7.	HFT of a Single Asset		141
	7.1	Stochastic Integral of Stationary Processes	142
		7.1.1 Ito-Riemann Sums and Their Limit	142
		7.1.2 Profit of HFT and Strong Ergodic Theorem	143
		7.1.3 Sharpe Ratio Test for HFT	147
	7.2	Two Examples	149
		7.2.1 The First Example	149
		7.2.2 The Second Example	151
8.	Bid, Ask and Trade Prices		155
9.	Financial Engineering		159
	9.1	Mathematical Finance	159
	9.2	Statistical Finance	162
	9.3	Behavioral Finance	163
	9.4	Computational Finance	164
10.	Debate and Future		167
References			171
Index			175

Foreword to *East China Normal University Scientific Reports*

Founded in 1951, as a result of merging of several old universities in Shanghai (including St. John's University founded in 1879), East China Normal University (ECNU) becomes one of the most prestigious universities in China. Our faculty members have acquired high academic reputations in a number of scientific research areas.

ECNU Scientific Reports, a book series edited by the editorial office of *Journal of ECNU* and published by World Scientific Publishing Co., present valuable results and significant progresses in scientific researches. We hope that this book series will serve as an international platform for scientific research, particularly inter-disciplinary research.

I am delighted to see the publication of the first volume, *High-Frequency Trading and Probability Theory*; and would like to sincerely congratulate the authors. I expect that more and more valuable volumes of the series will appear in the near future, so that the international scientific world will benefit from our research.

Qun Chen
President of East China Normal University
Shanghai, China

Preface

There have been quite a few publications on **High-Frequency Trading (HFT)**. However, few of them discussed the detailed strategies and algorithms of HFT. From our point of view, HFT is an application of high technology and mathematics to the financial market. Based on our practice in the Chinese futures market, we will show mainly three things in this book:

(1) Some facts and technical details of the IT system design for HFT, some of which have not been seen in literature until this point.
(2) HFT has the ergodic theory of stationary process as its mathematical background.
(3) A trader can use technical analysis to repeatedly trade one financial derivative and make statistical arbitrage according to the strong ergodic theorem.

Due to its remarkable profitability, the algorithms of HFT were considered highly confidential and have not been well-studied in the academic world. Just as in Chinese proverb "throwing out a brick in order to attract the others throwing gems", we hope that this book will attract more researchers to study the scientific background of HFT. We will concentrate our discussions of the examples of the Chinese futures market. However, the general principles apply to other markets as well.

We assume that our readers are aware of some basic calculus. The material of this book was discussed at the School of Finance and Statistics, East China Normal University three times in three semesters. We would like to thank Professor Shujin Wu, PhD students Si Bao, Shi Chen, Chang Liu, Shuai Wang and Yu Zhou for various insightful discussions and technical assistances. We particularly thank Ms. Xiaolin Guo who videotaped the first author's lectures which form the first part of this book, and thank Ms. Eileen Raney who assisted us to edit the original English manuscript. This work is partially supported by the 111 project (B14019) in China.

Zhaodong Wang and Weian Zheng
School of Finance and Statistics
East China Normal University
Shanghai 200241, China
December 2013

About the Authors

Zhaodong Wang obtained his PhD degree in Computer Science from Shanghai Jiao Tong University. He is an expert in high performance system design. He was the CEO of Shanghai Futures Information Technology Corp. Ltd. At that time, he designed the trading system for Shanghai Futures Exchange, China Finance Futures Exchange and many other systems in financial area. Later, he set up a hedge fund focus on high-frequency trading.

Weian Zheng is a State Endowed Professor at ECNU and Professor Emeriti at University of California, Irvine, USA. He had Docteur d'état ès Sciences Mathématiques degree from University of Strasbourg, France. He has a few known results in probability theory, including (with P. A. Meyer) the tightness criteria for laws of semi-martingales and (with T. J. Lyons) the forward–backward decomposition of symmetric Markov processes. With Y. Shen, he used probability theory to reduce the famous Monge–Kantorovich mass transfer problem to a boundary value problem of partial differential equation and solved this problem in the plane. He was born in Shanghai and grew up during the epoch of Chinese Cultural Revolution. He was unable to continue his study after the first year of middle school. He studied mainly by himself to become a mathematician.

Chapter 1

Introduction

In recent years, **high-frequency trading (HFT)** became an extremely hot topic in capital markets around the world. Normally, HFT has the following attributes:

- It is a kind of automatic trading strategy, which means that the trading is done by computer and there is no instance of human decision making during the buying or selling of any instrument. Some HFT systems allow traders to have some influence on it, but they can only change a few parameters of the strategy, and must allow the system to apply these parameters to the strategy making the trades.
- The traders keep positions for a very short time. In most cases, only several seconds or minutes. Some people believe that traders should not take any position over night, while others think it is allowed. But most agree that, HFT should not keep the main parts of the position to the second day.
- High-frequency traders only trade via electronic trading systems, not over-the-counter (OTC) markets, or any market still using outcries.
- It uses a high-speed connection to connect to the market, so as to retrieve high-frequency market data and place orders. In most cases, HFT requires the fastest access methods for a specified market, for example, the securities exchange place.
- Low latency is always very important for HFT. Various technologies are used for this target, including software optimization, hardware accelerators and dedicated network equipments. The time difference between an input message and corresponding order insert action has

been defined as the **internal response time**, which is a key benchmark of latency for HFT. Recent competition has reduced it to several microseconds.
- Unlike other low latency trading strategies, high-frequency traders normally calculate profit and Sharpe ratio on a daily basis, not just annually.

Automatic trading is a result of wide acceptance of electronic trading platforms. Compared with doing these strategies manually, automatic trading has a lot of advantages: accuracy, objectiveness, no emotion, and low cost. Therefore, it is becoming more and more popular. Any strategies that are without the feelings of traders can be converted to an automatic trading strategy. As the market grows, the trading costs become lower and lower, allowing for more high-frequency strategy to apply to the market. A trader cannot guarantee that his trades are all with positive return. However, if the strategy is correct and can be applied repeatedly in a statistically stationary way, then according to the strong ergodic theorem, the accumulated profit will be increasing at a stable rate. Therefore, all automatic traders may try to find some HFT algorithm for a large accumulated volume. So we can say that, HFT is the result of market and technological improvements.

At the end of the last century, electronic trading platforms have become the main trading media for the US and European exchanges. HFT was created by proprietary trading firms, hedge funds and investment banks. As HFT achieved higher profit rates and stable performances, more and more institutional investors started to use it. It became so popular that in 2009, 73% of the US equity trading volume was made by HFT firms, accounting for only 2% of 20,000 funds [15].

Like any new financial tools, the HFT comes with a considerable amount of controversy. Recall the discussion about the greedy usurer in the middle ages, and the consequence of South Sea Bubble on the London Stock Market, the cases are somewhat similar to what is happening for HFT.

In this book, we first discuss some classical program trading based on the Chinese futures market. The classical program trading is a trade of a basket of assets which is executed by a computer program (accurate to the level of microseconds) based on a predetermined algorithm. There have been essentially two reasons to use this type of program trading.

(a) When one desires to trade several stocks or futures at the same time (for example, when a mutual fund receives an influx of money, it will use that money to increase its holdings in the multiple stocks on which the fund is based);
(b) When one wishes to arbitrage temporary price discrepancies between related financial instruments, such as between an index and its constituent parts.

Probability theory and statistics are deeply involved in the two types of trading mentioned above, because of the existence of the time gap (although only relevant for a few seconds) between the exchange platform sending out the current bid-ask information and receiving traders' order. The main difficulty is that a high-frequency trader may take a loss if his orders have been only partially executed. Therefore, a successful trader should design an algorithm such that his profits may be related to an ergodic stationary process with a positive mean so that the strong ergodic theorem assures the accumulated profitability increasing in a stable manner.

Generally speaking, the more assets in one's basket to trade, the more risk that his order will be only partially executed. Therefore, it is wise to reduce the number of assets in the trading basket. A problem is raised naturally: can one trade only one asset repeatedly to take stable profits in HFT? In the second part of this book, we discuss some "new" (possibly known by some traders but unpublished due to its profitability) algorithms of HFT which is built on the ergodicity of stationary processes. We show that the technical indicators used in the market form a multidimensional stationary process, which may lead considerable statistical arbitrage in HFT. The technical indicators have a mathematical and statistical background. They work much better in HFT than in ordinary trading, which can be explained from the behavioral finance point of view (see Section 9.3).

Our method is also useful when one wants to trade a large quantity of assets. A direct trade order of a large quantity of an asset will affect the price by pushing it toward the undesired direction (for example, imagine what would happen if a trader bids several millions of shares of a stock — the price will jump). So if a trader knows the algorithms of HFT, he knows how to separate his orders to reduce the buying cost and get a better selling price.

Let us briefly describe here one asset HFT. The price $P(t)$ at time t of certain asset is a stochastic process in t. That is, for each fixed t, $P(t)$ is a random variable. Denote $p(t) = \log P(t)$ and $\Delta p(t)$ as its increment, i.e., $\Delta p(t) = p(t) - p(t - \delta)$ for some fixed positive δ. We will mainly use per 0.5 second high-frequency data, so our time unit is 0.5 second and $\delta = 1$ unit (0.5 second). $\{\Delta p(t)\}$ is called as the **logarithmic return** in finance, which can be considered as a **strongly stationary** process (see Sections 5.2 and 6.2) for certain heavily traded securities during the main trading hours. That will be our major hypothesis throughout the whole book. That hypothesis is in the common core of most popular mathematical models for security price. As we know, the more sophisticated a stochastic model might be, the more difficult to be verified in applications. Therefore, we select the simplest one, which can be (partially) tested by statistics.

We show that the basic technical indicators can be considered as (or transformed into) a multidimensional function

$$X(t) = f(\Delta p(s), t - q \leq s \leq t),$$

which depends only on the logarithmic returns in the past time interval $[t - q, t]$ of the length q. As a function of stationary process, $\{X(t)\}$ is also stationary. When we use an algorithm based on $\{X(t)\}$ to repeatedly trade one unit of the asset, then the instant logarithmic return will be $\{H(t-1)\Delta p(t)\}$, where $H(t-1) = 1$ if we have the asset or $H(t-1) = 0$ if we do not have the asset at time $t - 1$. These instant logarithmic returns form a stationary time series. Denote by $M(T)$ the cumulated logarithmic return by time T after deduction of the trading costs. Then, we can use the strong ergodic theorem to show that the mean logarithmic return $M(T)/T$ converge for large T. The limit will be a positive constant when the instant logarithmic return is ergodic and has positive mean after cost deduction. That is the mathematical background of HFT. An interesting remark is that when one uses 0.5 second as the time unit, four trading hours are equivalent to $T = 28,800$ (0.5 second), which is a quite large number in mathematical statistics.

The algorithm of HFT depends on the microstructure of the market. The trading regulations and transaction costs heavily affect the HFT algorithms.

Introduction

In this book, we use the Chinese futures market as an example. Nevertheless, most of our principles also apply to the other markets.

China has the fastest growing capital market in the world, especially after the release of the CSI 300 index futures contracts in 2010. Here is a figure on the growth of trading volume of CSI 300 futures in 2012 (based on public data from the China Financial Futures Exchange).

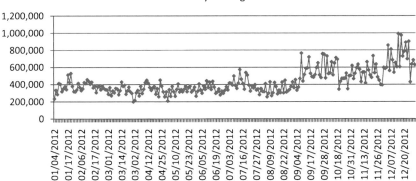

Here is a sub-table on top index futures contracts in 2012 from FIA [1], which illustrates the tremendous growth of Chinese index futures market.

Rank	Contract	2011	2012	% Change
1	E-mini S&P 500 Futures, CME	620,368,790	474,278,939	−23.50
2	RTS Futures, Moscow Exchange	377,845,640	321,031,540	−15.00
3	Euro Stoxx 50 Futures, Eurex	408,860,002	315,179,597	−22.90
4	Nikkei 225 Mini Futures, OSE	117,905,210	130,443,680	10.60
5	CSI 300 Futures, CFFEX	50,411,860	105,061,825	108.40

The Chinese capital market has electronic trading platform, which has no obstacle for automatic trading. However, the high trading costs and low liquidity made HFT unavailable until 2005 in the futures market. We cannot find any official statistics on the percentage of the trading volume made by HFT in China, but it is widely believed that at least 20% of futures market trading volume comes from HFT. On the other hand, because of the $T + 1$ rule (the shares cannot be sold on the same day that they have been purchased) in the Chinese stock market, HFT can rarely be used in cash markets in China. HFT is fast growing, and has not yet reached the peak in current Chinese derivative markets.

We discuss in Chapter 2 the microstructure of the Chinese futures markets, including trading, clearing, market data, risk control, etc. If you are quite familiar with the market microstructure of the Western exchanges, you can skip most of the sections of this chapter, and just read the last section, which describes the differences between the Chinese futures market and the Western markets. Based on our market microstructure understanding and simple mathematical tools, we can analyze classical HFT strategies, and their applications in the Chinese market, which are the topics of Chapter 3. After having a suitable strategy, the next problem is to design an IT system to implement; this is described in Chapter 4. In order to reach more readers from disciplines other than mathematics, we use Chapter 5 to introduce the basic concepts of probability and statistics, especially the ergodic theorem for stationary process. Mathematicians may skip that chapter, except for the last example of HFT based on stationary process. We use Chapter 6 to show that the technical analysis of financial market has a statistical foundation. Technical indicators are associated with stationary processes. Therefore, a trader can calculate statistics based on these stationary processes and the strong ergodic theorem that guarantee the observed relative frequencies will converge to the corresponding probabilities. Chapter 7 is used to describe the mathematical foundation for HFT of a single asset and gives two examples. We will show how to use the bid-ask spread and the last price to insert trade orders in Chapter 8. HFT can be considered an important application of financial engineering. Chapter 9 is an overview of financial engineering including a few related fields such as computational finance, mathematical finance, statistical finance and behavioral finance. Chapter 10 is used to discuss the future of HFT.

Chapter 2

Market Microstructure

Algorithms of high-frequency trading (HFT) are heavily based on the market. A profitable algorithm in one market may not work in another market. That is the reason why we need to study the market microstructure on which our algorithms are based.

2.1 Trading Products

Currently, there are four futures exchanges in China, **China Financial Futures Exchange (CFFEX), Shanghai Futures Exchange (SHFE), Dalian Commodity Exchange (DCE)** and **Zhengzhou Commodity Exchange (CZCE)**.

CFFEX is the only futures exchange focusing on financial products in China. It was set up in 2006, and launched its first financial product, **CSI 300 futures (Symbol IF)**, in 2010. CFFEX is growing so fast that it has become the largest futures exchange in China according to trading turnover. The CSI 300 is a stock index based on 300 stocks, including most of the important stocks in the Shanghai Stock Exchange and the Shenzhen Stock Exchange. It is a key benchmark of the performance of all listed companies in China. The rules of the CSI 300 [8] are very similar to those of the Hang Seng Index Futures in HKEX. The contract months include spot, next month and next two quarter months. The maturity date of each spot contract is the third Friday of that month. The trading hour is from 09:15 to 11:30 and from 13:00 to 15:15 (China standard time), which covers the trading hour of the cash market with a leading and a tailing quarter extension. On the maturing dates, the trading will end at 15:00. The opening auction is from 09:10 to 09:14. The final settlement price is the average CSI 300 index price of every 5 minutes in the afternoon of the last trading day. The contract multiplier is CNY 300 per index point. According to recent prices, the contract value is about CNY 700,000, a little higher than US$110,000 according to the current exchange rate. The tick of IF is 0.2, hence the tick value is CNY $300 \times 0.2 =$ CNY 60, near US$10.

All the other three exchanges are focused on various commodities, including most of the important industrial and agricultural products. The following table is a list of all futures products in China at the end of 2013 (tick is in CNY) [9, 10, 25]:

Category	Product	Exchange	Symbol	Contract Months	Multiplier	Unit	Tick
Financial	CSI 300 index	CFFEX	IF	Current, next and next two quarter months	300	Point	0.2
Financial	Five-year treasury bond	CFFEX	TF	Three recent quarter months	10,000	100 CNY	0.002
Metal	Copper	SHFE	cu	Next 12 months	5	Ton	10
Metal	Aluminum	SHFE	al	Next 12 months	5	Ton	5
Metal	Zinc	SHFE	zn	Next 12 months	5	Ton	5
Metal	Lead	SHFE	pb	Next 12 months	25	Ton	5
Metal	Steel rebar	SHFE	rb	Next 12 months	10	Ton	1
Metal	Steel wire rod	SHFE	wr	Next 12 months	10	Ton	1
Metal	Gold	SHFE	au	Next 12 months	1,000	Gram	0.01
Metal	Silver	SHFE	ag	Next 12 months	15	kg	1
Agriculture	Nature rubber	SHFE	ru	Next 12 months except Feb and Dec	10	Ton	5

(*Continued*)

(*Continued*)

Category	Product	Exchange	Symbol	Contract Months	Multiplier	Unit	Tick
Energy	Fuel oil	SHFE	fu	Next 12 months except spring festival month	50	Ton	1
Energy	Bitumen	SHFE	bu	Next 6 months and all quarterly month within 24 months	10	Ton	2
Agriculture	Soybean No. 1	DCE	a	Next Jan, Mar, May, July, Sep and Nov	10	Ton	1
Agriculture	Soybean No. 2	DCE	b	Next Jan, Mar, May, July, Sep and Nov	10	Ton	1
Agriculture	Soybean meal	DCE	m	Next Jan, Mar, May, July, Aug, Sep, Nov and Dec	10	Ton	1
Agriculture	Soybean oil	DCE	y	Next Jan, Mar, May, July, Aug, Sep, Nov and Dec	10	Ton	2
Agriculture	Corn	DCE	c	Next Jan, Mar, May, July, Sep and Nov	10	Ton	1
Agriculture	RBD palmolein	DCE	p	Next 12 months	10	Ton	2
Agriculture	Egg	DCE	jd	Next 12 months	10	500 kg	1
Energy	Coke	DCE	j	Next 12 months	100	Ton	1
Energy	Coking coal	DCE	jm	Next 12 months	60	Ton	1
Chemical	LLDPE	DCE	l	Next 12 months	5	Ton	5
Chemical	PVC	DCE	v	Next 12 months	5	Ton	5
Chemical	Fiberboard	DCE	fb	Next 12 months	500	Piece	0.05
Chemical	Blockboard	DCE	bb	Next 12 months	500	Piece	0.05
Mineral	Iron ore	DCE	i	Next 12 months	100	Ton	1

(*Continued*)

(*Continued*)

Category	Product	Exchange	Symbol	Contract Months	Multiplier	Unit	Tick
Agriculture	Common wheat	CZCE	PM	Next Jan, Mar, May, July, Sep and Nov	50	Ton	1
Agriculture	Strong gluten wheat	CZCE	WH	Next Jan, Mar, May, July, Sep and Nov	20	Ton	1
Agriculture	Cotton	CZCE	CF	Next Jan, Mar, May, July, Sep and Nov	5	Ton	5
Agriculture	Sugar	CZCE	SR	Next Jan, Mar, May, July, Sep and Nov	10	Ton	1
Agriculture	Rapeseed	CZCE	RS	Next July, Aug, Sep and Oct	10	Ton	1
Agriculture	Rapeseed meal	CZCE	RM	Next Jan, Mar, May, July, Sep and Nov	10	Ton	1
Agriculture	Rapeseed oil	CZCE	OI	Next Jan, Mar, May, July, Sep and Nov	10	Ton	2
Agriculture	Early rice	CZCE	RI	Next Jan, Mar, May, July, Sep and Nov	20	Ton	1
Agriculture	Japonica	CZCE	JR	Next Jan, Mar, May, July, Sep and Nov	20	Ton	1
Chemical	PTA	CZCE	TA	Next 12 months	5	Ton	2
Chemical	Methanol	CZCE	ME	Next 12 months	50	Ton	1
Energy	Thermal coal	CECZ	TC	Next 12 months	200	Ton	0.2
Other	Glass	CZCE	FG	Next 12 months	20	Ton	1

The trading hours for the above commodity futures are the same, from 09:00 to 11:30, and from 13:30 to 15:00. From 08:55 to 08:59 is the open auction time (all in China standard time). There is a big pressure for exchanges to open night sessions from traders, as most of those products have overseas markets, which trade in the night according to China standard time. Some products have night trading sessions, such as gold and silver (from 21:00 to 02:30), as well as copper, aluminum, lead and zinc (from 21:00 to 01:00).

The contract ID of Chinese futures is quite different from the Western customs. There is no traditional usage of a letter to represent a month in China. Chinese use two digits to represent months instead, and the year is put before the month. For example, the copper contract for February 2014 has ID cu1402. Most Chinese futures contracts have this kind of ID, except CZCE, which uses one digit to represent the year. Hence, the sugar contract for May 2014 has the ID SR405.

All the commodity futures in China use physical delivery. Different products have different delivery dates in the spot month. Fixed premium is introduced to handle the differences in product quality and delivery place. Similar to Western markets, there is a low delivery volume compared with the trading volume.

There are some other exchanges in China which are trading some futures-like products. The most important one is the Shanghai Gold Exchange (SGE). They provide a carrying trading product, including gold and silver, named as T + D contracts. The rule is similar to the Chinese Gold and Silver Exchange Society in Hong Kong. Buyers and sellers can trade during the trading hours, and then some of them may submit a request for delivery at the end of the day. Some requests can be fulfilled by matching the requests from the other direction. If there are some buyers remaining and not enough sellers, all the sellers should pay a carrying fee to all the buyers, and vice versa. If these requests are exactly matched, no side will pay the carrying fee. This is a kind of cash market trading, where everyone can delay the delivery time by paying the carrying fee. All the other parts of T + D contracts are very similar to futures, including the trading model, risk control and so on. SGE also provides night trading session to cover the trading hours in the US.

2.2 Trading Model

Similar to Western markets, Chinese futures markets have two main trading models: (i) continuous trading and (ii) auction. We will describe in detail the trading process of each model.

2.2.1 Continuous Trading

Continuous trading is the main trading model of the market, which lasts most of the trading hours. During the continuous trading session, clients will submit their buying or selling indication via orders to the exchanges, and the exchanges will use a centralized order book to keep all the orders and find a suitable match of these orders and then make the trade. The following diagram briefs the process of handling an order in the exchange trading system.

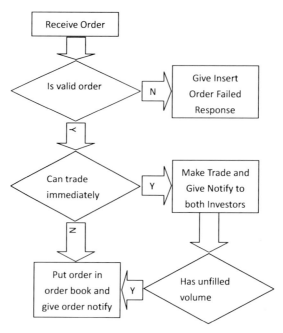

There are three main parts of the process in this diagram: (i) order validation, (ii) order book management and (iii) matching process.

The order validation has various different rules that should be checked, including the following:

- Field check determines whether all the required fields are properly set, and all the information in the order is consistent.
- Reference check determines whether all the reference information is correct, including information specifying instrument, member, client reference and etc. It should be noted that in China, every order submitted

to the exchange should set a client reference field, to specify a client before trading.
- Risk check determines whether all the anti-risk rules are satisfied. The significant parts of risk check are to check position and money. China exchanges should keep all the positions for each client, and should check whether the position limit of a member or a client is exceeded while handling open orders, and check that there is enough position while handling close orders. It is another difference between the Chinese market and Western markets that clients should specify either open or close in their orders, which means that China uses gross position for clients instead of net position. To check money, China exchanges should calculate the maximum margin required in case the order is totally filled, and check whether the member has enough money to cover it for open orders, as close orders will free margin, not freeze margin. It is different with the position check — money check is for members, not for clients, as the exchange does not know the exact monetary information of each client, only the members know about it. Therefore, money checks for each client should be done on the member side, which is in the member side trading system. Risk check is a large topic, which will be discussed further later.

In the order management of an exchange, the exchange will maintain two order books for each instrument: the buy order book and the sell order book. Each order, which is not totally filled and not cancelled, will be put into the corresponding order book based on its priority. In most cases, the priority is price-time. This means the prices must be looked at first, if the prices are the same, then order insert time must be compared. In the buy order book, the higher the order price, the higher the priority. In the sell order book, the lower the order price, the higher the priority. When compared with time, the earlier the higher. There is one exception in SHFE, however, which is another priority between price and time — whether or not it is a closed previous position order on price limit. First, we should mention that in SHFE, all the positions are split into two parts: today's position and previous positions. Any position opened before today is a previous position, and all positions opened today are today's position. When the client wants to close a position, he or she should specify whether to close today's position or the previous position. All the other three futures exchanges do not have similar arrangements. The unique priority in SHFE indicates that if somebody submits a closed previous position order with the price in price limit, it will be a higher priority than an open or close today order at the same price. The order price should be the upper price limit for the buy order and the lower price limit for the sell order in order to be higher priority. The following is an example of the order book in this rule. It is a buy order book, and listed from highest priority to lower. The upper price limit is 64,000.

Price	Offset Flag	Insert Time	Comment
64,000	Previous	10:02:05	Closed previous position in price limit gains higher priority.
64,000	Close previous	10:03:11	Same with previous, so as to compare time.
64,000	Open	10:01:07	Not closed past position, then compare the time.
64,000	Close today	10:01:13	Same priority for open and closed today, so compare time.
64,000	Open	10:01:34	
64,000	Close today	10:01:44	
63,950	Open	10:00:05	Lower price, hence lower priority.
63,950	Close previous	10:00:07	Not in price limit, hence no special priority for closed past position.
63,950	Close today	10:01:11	

The matching process is to match buy orders and sell orders to make a trade. If the buyer's price is higher than or equal to a seller's price, then the trade can be made. The exchange trading system will try to match any incoming order with the highest priority order in the opposite order book. If a match is possible, a trade with the minimum of the two opposite orders is made. If the order in the order book is totally filled, it will be removed from the order book, and if the incoming order still has remaining volume, the trading system will try to match it again, until the incoming order is fully filled, or it cannot make any further trades. If there are still remaining volumes in the incoming order, it will be put into the order book of its own direction. Now let us discuss the trade price. In most Western exchanges, the trade price is the first price, which is the price of the order in the order book. All Chinese futures markets, however, use another rule, **the Middle**

Price Rule. That is, the middle price of the buy price, sell price and last price. The last price is the price of the last trade. We know that any prices between buy price and sell price are reasonable for both sides. The middle price finds a price in this region which is mostly closer to last price. Using this rule will minimize the movement of last price.

2.2.2 Auction

The auction model is used during the opening session. Opening session is used to find a suitable open price for each instrument. All the clients may submit orders during the auction session, but no trade will be made immediately. At the end of an auction session, all the orders are matched together. Similar to continuous trading, the trading system will validate all the orders before accepting them. After validation, all the orders will be put in order books till the end of the auction. The core idea of an auction is to draw the supply–demand diagram and try to find the intersection that indicates the suitable price and trade volume. In the real trading system, all the prices and volumes are discrete; therefore, the trading process is somewhat different. The auction process can be divided into two parts, price determination and trade allocation.

To determine the trade price of auction, trading system will first gather all the orders in auction session, and make a table based on the total order volume of buy orders and sell orders at each price. Here is an example, assuming that the tick is 0.2.

Price	Buy Volume	Sell Volume
2,100.0	0	9
2,099.8	0	7
2,099.6	0	5
2,099.4	0	1
2,099.2	0	2
2,099.0	0	0
2,098.8	1	4
2,098.6	3	6
2,098.4	1	3
2,098.2	1	3
2,098.0	0	1

(*Continued*)

	(Continued)	
Price	Buy Volume	Sell Volume
2,097.8	3	3
2,097.6	3	0
2,097.4	4	1
2,097.2	1	0
2,097.0	0	0
2,096.8	2	0
2,096.6	3	0
2,096.4	6	0

In the above table, each volume number is the sum of all orders in this price and this direction. We should note that all prices have to be listed even if there is no order on that price, for example, 2,099.0 and 2,097.0.

The second step is to calculate the accumulative buy volume and accumulative sell volume. The accumulative buy volume is the sum of all the buy volumes from highest price to current price. The accumulative sell volume is the sum of all the sell volumes from lowest price to current price. Here is the table of the previous example.

Price	Buy Volume	Sell Volume	Accumulative Buy Volume	Accumulative Sell Volume
2,100.0	0	9	0	45
2,099.8	0	7	0	36
2,099.6	0	5	0	29
2,099.4	0	1	0	24
2,099.2	0	2	0	23
2,099.0	0	0	0	21
2,098.8	1	4	1	21
2,098.6	3	6	4	17
2,098.4	1	3	5	11
2,098.2	1	3	6	8
2,098.0	0	1	6	5
2,097.8	3	3	9	4
2,097.6	3	0	12	1
2,097.4	4	1	16	1
2,097.2	1	0	17	0
2,097.0	0	0	17	0
2,096.8	2	0	19	0
2,096.6	3	0	22	0
2,096.4	6	0	28	0

The third step is to calculate the possible trade volume on each price, which is the minimum of the accumulative buy volume and the accumulative sell volume. Here is the table of the previous example.

Price	Buy Volume	Sell Volume	Accumulative Buy Volume	Accumulative Sell Volume	Trade Volume
2,100.0	0	9	0	45	0
2,099.8	0	7	0	36	0
2,099.6	0	5	0	29	0
2,099.4	0	1	0	24	0
2,099.2	0	2	0	23	0
2,099.0	0	0	0	21	0
2,098.8	1	4	1	21	1
2,098.6	3	6	4	17	4
2,098.4	1	3	5	11	5
2,098.2	1	3	6	8	6
2,098.0	0	1	6	5	5
2,097.8	3	3	9	4	4
2,097.6	3	0	12	1	1
2,097.4	4	1	16	1	1
2,097.2	1	0	17	0	0
2,097.0	0	0	17	0	0
2,096.8	2	0	19	0	0
2,096.6	3	0	22	0	0
2,096.4	6	0	28	0	0

In the above table, we can find there is one price, 2,098.2, whose corresponding trade volume reaches the maximum value, 6. Therefore, 2,098.2 is the trade price of this auction. This is the **Maximum Trade Volume Rule**.

If there are several prices that have the same maximum trade volume, the trading system should try to find one price from them for the final trade price. Then trading system will do the fourth step, calculate the remaining volume on each price, which is determined by the maximum of the accumulative buy volume and the accumulative sell volume minus trade volume. The remaining volume indicates how many volumes will remain unfilled in this price if we trade at this price. The following is an example which has multiple maximum trade volumes.

Price	Buy Volume	Sell Volume	Accumulative Buy Volume	Accumulative Sell Volume	Trade Volume	Remain Volume
2,100.0	0	9	0	45	0	45
2,099.8	0	7	0	36	0	36
2,099.6	0	5	0	29	0	29
2,099.4	0	1	0	24	0	24
2,099.2	0	2	0	23	0	23
2,099.0	0	0	0	21	0	21
2,098.8	1	4	1	21	1	20
2,098.6	3	6	4	17	4	13
2,098.4	1	3	5	11	5	6
2,098.2	0	3	5	8	5	3
2,098.0	0	1	5	5	5	0
2,097.8	3	3	8	4	4	4
2,097.6	3	0	11	1	1	10
2,097.4	4	1	15	1	1	14
2,097.2	1	0	16	0	0	16
2,097.0	0	0	16	0	0	16
2,096.8	2	0	18	0	0	18
2,096.6	3	0	21	0	0	21
2,096.4	6	0	27	0	0	27

In this example, we can find three prices, 2,098.4, 2,098.2 and 2,098.0 that share the same maximum trade volume 5. Among them, "2,098.0" has the minimum remaining volume 0. The trading system will select 2,098.0 as the trade price for this auction. This is the **Minimum Remain Volume Rule**.

If there are still several prices that have the same remaining volume, the rest of the auction algorithms are different in different exchanges. We will describe the algorithms for CFFEX and SHFE in detail. The trading system will extend the table and calculate the difference between the accumulative buy volume and the accumulative sell volume. Here is an example that has a multiple of minimum remaining volume.

Price	Buy Volume	Sell Volume	Accumulative Buy Volume	Accumulative Sell Volume	Trade Volume	Remain Volume	Accumulated Difference
2,100.0	0	9	0	45	0	45	−45
2,099.8	0	7	0	36	0	36	−36

(*Continued*)

Market Microstructure

(Continued)

Price	Buy Volume	Sell Volume	Accumulative Buy Volume	Accumulative Sell Volume	Trade Volume	Remain Volume	Accumulated Difference
2,099.6	0	5	0	29	0	29	−29
2,099.4	0	1	0	24	0	24	−24
2,099.2	0	2	0	23	0	23	−23
2,099.0	0	0	0	21	0	21	−21
2,098.8	1	4	1	21	1	20	−20
2,098.6	3	11	4	17	4	13	−13
2,098.4	1	0	5	6	5	1	−1
2,098.2	0	0	5	6	5	1	−1
2,098.0	0	2	5	6	5	1	−1
2,097.8	3	3	8	4	4	4	4
2,097.6	3	0	11	1	1	10	10
2,097.4	4	1	15	1	1	14	14
2,097.2	1	0	16	0	0	16	16
2,097.0	0	0	16	0	0	16	16
2,096.8	2	0	18	0	0	18	18
2,096.6	3	0	21	0	0	21	21
2,096.4	6	0	27	0	0	27	27

We can find three prices, 2,098.4, 2,098.2 and 2,098.0, which share the same minimum remaining volume: (1) the trading system will check the accumulated difference, which is monotonously increasing, find the price most close to 0, which is 2,098.0 and 2,097.8. We can find that of the three trade price candidates, 2,098.0 is most close to this zero intersection. Then, 2,098.0 is the trade price for this auction. If there are still multiple prices according to this rule, we can prove that each price is a proper price for this auction. A proper price occurs if one trades in this price, all the buy orders with higher prices and sell orders with lower prices must be totally filled, at most, only one side of orders at this price may not be totally filled. The illustration will not be included in this book.

If there are still multiple prices according to this rule, the trading system will select one price which is closest to the last price as the final trade price for this auction. Below are two kinds of examples which may have multiple prices according to this rule.

This example is based on multiple zeroes in accumulated difference.

Price	Buy Volume	Sell Volume	Accumulative Buy Volume	Accumulative Sell Volume	Trade Volume	Remain Volume	Accumulated Difference
2,100.0	0	9	0	45	0	45	−45
2,099.8	0	7	0	36	0	36	−36
2,099.6	0	5	0	29	0	29	−29
2,099.4	0	1	0	24	0	24	−24
2,099.2	0	2	0	23	0	23	−23
2,099.0	0	0	0	21	0	21	−21
2,098.8	1	4	1	21	1	20	−20
2,098.6	3	12	4	17	4	13	−13
2,098.4	1	0	5	5	5	0	0
2,098.2	0	0	5	5	5	0	0
2,098.0	0	1	5	5	5	0	0
2,097.8	3	3	8	4	4	4	4
2,097.6	3	0	11	1	1	10	10
2,097.4	4	1	15	1	1	14	14
2,097.2	1	0	16	0	0	16	16
2,097.0	0	0	16	0	0	16	16
2,096.8	2	0	18	0	0	18	18
2,096.6	3	0	21	0	0	21	21
2,096.4	6	0	27	0	0	27	27

This example shows that both prices near the zero intersection have the same minimum remaining volume.

Price	Buy Volume	Sell Volume	Accumulative Buy Volume	Accumulative Sell Volume	Trade Volume	Remain Volume	Accumulated Difference
2,100.0	0	9	0	45	0	45	−45
2,099.8	0	7	0	36	0	36	−36
2,099.6	0	5	0	29	0	29	−29
2,099.4	0	1	0	24	0	24	−24
2,099.2	0	2	0	23	0	23	−23
2,099.0	0	0	0	21	0	21	−21
2,098.8	1	4	1	21	1	20	−20
2,098.6	3	11	4	17	4	13	−13
2,098.4	1	0	5	6	5	1	−1
2,098.2	0	1	5	6	5	1	−1
2,098.0	1	1	6	5	5	1	1
2,097.8	2	3	8	4	4	4	4
2,097.6	3	0	11	1	1	10	10

(*Continued*)

(Continued)

Price	Buy Volume	Sell Volume	Accumulative Buy Volume	Accumulative Sell Volume	Trade Volume	Remain Volume	Accumulated Difference
2,097.4	4	1	15	1	1	14	14
2,097.2	1	0	16	0	0	16	16
2,097.0	0	0	16	0	0	16	16
2,096.8	2	0	18	0	0	18	18
2,096.6	3	0	21	0	0	21	21
2,096.4	6	0	27	0	0	27	27

After determining the trade price, the trade volume is detected, and the trading system will allocate trades to related orders. All the buy orders and sell orders will be queued with the same order in continuous trading, and corresponding volumes will be selected to trade.

Unlike most Western exchanges, all Chinese futures markets will have a small time of pause after the auction sessions, and then start the continuous session. This is mainly due to historical reasons. At the end of auction, it may take some time to disseminate all the trades and market data in narrow band networks at that exact time.

2.3 Market Data Information

During the trading time, the exchange trading system will disseminate market data to the whole market. Here is a list of key fields of level 1 market data, which are provided by all the futures exchanges in China.

Field Name	Description
Instrument ID	Reference to an instrument.
Update Time	The time stamp of the market data.
Last Price	The price of the last trade of this instrument. If no trade was made today, the trade of yesterday, or even earlier will be used. If no trade was made for this instrument at all, reference price will be used. Reference price will be determined by the exchange before the launching of each instrument.

(Continued)

Field Name	Description
	(*Continued*)
Bid Price 1	The highest order price in the buy order book. It can be of no value if there is no order in buy order book.
Bid Volume 1	The sum of the remaining volumes of all buying orders in Bid Price 1. It can be zero if no order is in buy order book.
Ask Price 1	The lowest order price in the sell order book. It can be of no value if no order is in sell order book.
Ask Volume 1	The sum of the remaining volumes of all selling orders in Ask Volume 1. It can be zero if no order is in sell order book.
Open Price	The price of the first trade today. It can be of no value if no trade is made today.
Close Price	The Last Price at the close time today. It can be of no value if it is before close time.
Settlement Price	The settlement price of today. It can be of no value if it has not reached the close time, or settlement price has not been calculated up to now.
Previous Close Price	The Close Price of last trading day.
Previous Settlement Price	The Settlement Price of last trading day.
Highest Price	The highest Last Price today.
Lowest Price	The lowest Last Price today.
Trade Volume	The total volume of all trades today.
Turnover	The total trade value of all trades today.
Open Interest	The sum of the positions of all the clients now.

We shall notice that the Open Interest in the Chinese futures market is always reflecting the figure in real time, not the figure at the end of the last trading day. As client references will be provided for each order, the exchange can calculate positions for each client, hence, it is easy to sum to a total Open Interest during the trading hour.

There are two ways to calculate Trade Volume, one is only using the sum of one side of the trade, and the other is using the sum of both sides of the trade. We call the first one as uniside market data and the second one as dualside market data. The Trade Volume in dualside market data is always twice the Trade Volume in uniside market data. Field Turnover and Open Interest are the same. In the Chinese futures market, CFFEX is using uniside market data, whereas all the other futures exchanges are using dualside market data.

All Chinese futures exchanges provide a snapshot of market data instead of continuous market data. Chinese exchanges will **provide market data every half second**. If there is more than one activity during this half second, all these activities will be combined and only one set of market data describing the situation at the end of this half second will be disseminated to the market. If nothing happened during this half second, exchanges may not disseminate any market data for this instrument. It is also for historical reasons for exchanges to do so, as the bandwidth was very narrow a few years ago, and by using snapshot market data, bandwidth can easily be estimated.

CZCE is the only futures exchange in China that provides market data during the auction session. In the auction session, CZCE will calculate the trading result based on the current orders continuously. If no trade will be made, the market data will be the same as the market data during the continuous session. If some trades can be made, the Bid Price and Ask Price will be the trade price of the auction session up to now, and the Bid Volume will be the sum of all the orders volumes whose price is greater than or equal to this trade price, the Ask Volume will be the sum of the orders volumes whose price is less than or equal to this trade price.

All the Chinese futures exchanges will provide market data to members and market data vendors. The basic level 1 market data is the same. Some exchanges, including CFFEX and DCE, will provide level 2 market data to some selected data vendors. Members will get level 1 market data for free. Market data vendors should pay for the market data, and hence have the right to disseminate the market data. Level 2 market data will be more expensive than normal level 1 data. Because it may take a longer time to get market data from market data vendors, HFT traders normally get market data directly from exchanges or from members. But market data vendors will provide terminals which can view past market data and conduct analysis easily, which is still very important for manual operations. Market data vendors may deduct some other fields based on the basic market data provided by exchanges; however, they may only be guessing. Therefore, one should take careful note of which fields are from exchanges.

2.4 Trading Interface

Trading interface is extremely important for automatic trading. Clients should get real time market data, insert order, cancel order and get all the responses via their trading interface. There are two kinds of trading interfaces, exchange trading interface and member trading interface. All the exchanges will provide some exchange trading interface which allow members to access the market. If the client can get the permit from a member for Direct Market Access (DMA), he may use the exchange trading interface directly. If the client cannot use DMA, he may use the trading interface provided by the member trading system, which may have the same functions as the exchange trading interface, but with a lower speed.

In the Chinese futures market, all four exchanges provide their trading interface via C++ API now. In the beginning, the trading interfaces were quite different among these exchanges. In 2005, a standard named "Futures trading data exchange protocol (FTD)" [8] was released. Since then, all the exchanges have provided their trading interfaces based on the protocol with extensions of their own special business. Therefore, the concepts of all these trading interfaces are the same, and the processes are similar. We will describe the mechanism of FTD below, and clients can read the Application Programming Interface (API) documents from exchanges based on this understanding.

The most important concept in FTD is the communication model. A communication model is the means of coordination between two peers: the member side system and the exchange side system. There are three kinds of communication models in FTD, dialogue model, private model and broadcast model.

Dialogue model is the typical client/server model. In this model, a member side system submits a request to the exchange side system, and the exchange side system gives one or more responses corresponding to it. A request reference should be used in each request to identify all the responses to its corresponding request. There is no recovery mechanism for the dialogue model. If the member side system submits a request, and the connection is lost before any response is received, he will not know whether his request has reached exchange or not. In FTD, a dialogue model is used to submit orders and request queries. While using a dialogue model, you cannot assume that you can get response.

A private model is a reliable one-way communication model from an exchange to a member. Exchanges will keep a private message flow for each member, and each message will be marked by a continuous increasing sequence number. After connection, exchanges will always send all the messages in this flow to members in the correct sequence. If, on the member side, one finds there is a gap in the flow, he may assume that there may be some messages that have been lost during communication, and he should reconnect to the exchange and give the last received sequence number. Then, the exchange will resend all the messages with a higher sequence number in order. In FTD, a private model is used for order notify and trade notify information. And you can assume that you will get all these messages without any loss or disorder.

A broadcast model is another reliable one-way communication model from exchange to member. It has the same mechanism as a private model. The only difference is that the exchange will keep only one broadcast message flow for all members, not different flows for different members. In FTD, broadcast model is used primarily for market data, exchange notification and trading segment notification.

In order to know whether an order has been accepted by the exchange side system, member side systems need a proper mechanism. This mechanism is an order reference management system. There are two kinds of order references used as a primary key for all the orders. The first one is order system ID (OrderSysID), which is assigned by exchange side systems when it is first accepted. It is unique in the exchange every day. The second one is order local ID (OrderLocalID), which is assigned by the member side system. Each member side system should ensure that OrderLocalID and UserID (the reference for the member seat) compose another primary key of orders. Normally, member side systems will use a continuous sequence for all the orders from one member seat. When there is any problem on the connection between a member side system and an exchange side system, it will be reconnected, and the exchange side system will give the maximum OrderLocalID it has accepted to the member side system. By this way, the member side system will know whether its orders are accepted if it has not received any response or notification. Member side systems should use an increasing sequence starting from this maximum OrderLocalID add 1 for future OrderLocalID. Member side system can also resend all the orders if it is unsure about whether the orders have been accepted or not because a duplicate OrderLocalID will cause the order to be rejected.

We will demonstrate some key communication processes using a sequence diagram. All the arrows in blue are in a dialogue model, yellow in private model and red in broadcast model.

First, we will give a sequence diagram showing insertion of an order with no trade made immediately.

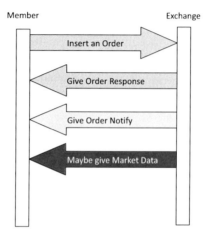

The next sequence diagram demonstrates inserting an order and making a trade immediately.

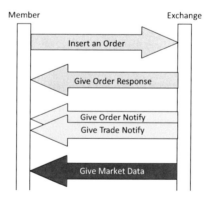

The third diagram shows how a trade is made on an order which has been inserted into the order book previously.

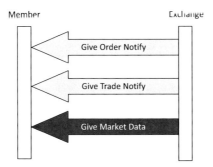

The fourth diagram shows how to cancel an order.

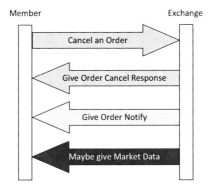

The last diagram shows an order which is rejected by exchange.

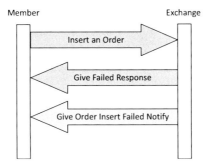

We may notice that there is both a response in the dialogue model and notification in the private model in most cases. Clients can select part of the information to handle. It is suggested that clients always handle notifications in the private model, and do not handle responses in dialogue model. Because all communication in private mode is in a synchronized model, responses in a dialogue model are more like a synchronized model, which existed mostly for historical reasons. It is important to realize that if a member side system gets the private message flow from the beginning of the day, it will have enough information to understand the current situation.

All four futures exchanges provide independent market data API which uses a dedicated connection. When clients make this connection, they specify the resume method for market data according to their requirement. There are three kinds of resume methods: from the beginning, from the break point and from now. If a client wants to resume from the beginning, all market data from the beginning of that day will be sent to the client. If a client selects resume from the break point, the exchange system will start from the point at which the client lost connection, just like in the private model. If it is the first time connecting that day, these two methods are the same. Most clients will select the last method, resume from now. This signals the exchange to send a snapshot of all the instruments at first, and then send any change in them afterwards. For trading purpose, clients can always use this method. The other two are mostly for usage by the market data vendor, who has the requirements to get all market data even if there is some interruption of connection.

There are some other important mechanisms in the APIs of the Chinese futures exchanges. For example, all these APIs have heartbeats, which can be used to detect any problem if the connection goes through the same wide area network. And all the exchanges will have some restriction on flow controls. Some controls are based on how many messages can be submitted to an exchange in one second, some are based on the maximum pending request count. The exact figure of these flow controls is often changed by exchange.

There are several key member side system vendors in China, including the Shanghai Futures Information Tech (SFIT), Sungard, Henshen and Yisheng. Among them, SFIT's system, CTP, is the most popular for program traders. CTP's trading interface is very similar to the exchange trading interface. The benefit of using CTP is that you can connect to one interface and trade in all four Chinese futures exchanges. Though the CTP system is very fast, it is still slower than DMA, so clients may use it for strategies not so critical for speed.

In order to get the fastest speed, another important issue is the location of a clients system. Presently, almost all the futures exchanges in China provide colocation service to members, and clients can get the spaces from their respective member. Some exchanges have independent buildings, and any offices in these buildings can connect to corresponding exchanges quickly. It has a similar effect as colocation.

There has been much debate about whether exchanges should provide trading interfaces in API, or in network protocol. It will be much easier for clients to write the system using API, and exchanges may put some important client side code in API, which can ensure the correct usage of communication. Additionally, exchanges can put some restrictions to secure the exchange side system. But programming in network protocol gives much more flexibility to clients, and they can use the latest technology to improve their system. Network protocol has the added benefit of no restrictions on hardware and operation systems. There are also some debates about the unfairness of using such kinds of technology. It is very hard to determine and will probably be discussed further more in the future.

2.5 Risk Control

Risk control is always a core problem in the derivative market. The Chinese futures market is a standard futures market, using almost the same rules as Western futures markets. The only difference is that the Chinese futures exchanges will do pre-trading risk control, including position checks and money checks.

The first important risk control rule in the futures market is the **Mark to Market rule**. It assumes that there is a suitable price for each instrument at the end of day, called the **settlement price**, and all buyers and sellers should calculate the profit or loss according to this price. All loss should be paid to the **Centralized Counterparty (CCP)** immediately, and the CCP must pay for any profit at this time too. For example, consider two sides that have made a trade on cu1401, for 1 lot, where the price is 50,000 and the settlement price is 50,200. The buyer will have a profit of 200 per ton, and if the multiplier of cu is 5, the buyer has a profit of 1,000, and the seller has a loss of 1,000. The seller should give 1,000 to the CCP, and the CCP will give 1,000 to the buyer at the end of day. This 1,000 and $-1,000$ are called the **position profit**. If on the next day, the settlement price is 50,300. The buyer will have a profit of 100 per ton, because the price increased from 50,200 to 50,300, which means 500 for this lot, and the seller loses 500. The seller will pay another 500 to the CCP, and the buyer will get it from the CCP. If, on the third day, both sides agree to close this position at 50,100, the buyer will lose 1,000. The price will have moved from 50,300 to 50,100, and the seller will win 1,000. That money will also be paid to and from the CCP. That kind of money is called close profit. In total, the buyer will make 500 as profit, and the seller will make 500 as loss. Considering that they opened at the price of 50,000, and closed at 50,100, the total result of the Mark to Market rule is just the same as calculating the difference between open and close price.

All instruments of futures should use the Mark to Market rule in everyday clearing, so that the counterparty risk for a long-term forward contract will be split on a daily basis. No one will suffer from a large price movement because the counterparty will not be willing to execute the contract. When the large price movement has been split into many smaller price movements and spread over multiple days, the counterparty will not have a very large loss on one day, and in most cases, they will keep the contract. If they do not pay for their loss one day, it will be considered a default, and a punishment will be introduced by the CCP. This is a part of the margin rule described later. If this punishment is always larger than their daily loss, they will not be likely to default according to the Hypothesis of Economic Man.

The margin rule is that both sides of the trade freeze a set amount of money in the CCP to cover potential losses of the next day. This frozen money is called a margin. The CCP is playing a balance on margin, not too low to have to risk the possibility of not having enough money to cover the potential loss, and not too high to expel investors. In the Chinese futures market, the entire margin is calculated by a margin rate. The formula is as follows:

margin = lot count × settlement price × contract multiplier × margin rate.

For example, if the margin rate for cu1401 is 7%, settlement price is 50,100, and you have 2 lots of positions, regardless you are long or short, the margin required is

$$2 \times 50,100 \times 5 \times 7\% = 35,070.$$

The Chinese futures exchanges, acting as the CCP, will change margin rates according to some predefined rules. Normally, when a contract comes near expiration, the margin rate will increase. For products that are also traded abroad, if it is the last trading day before a long holiday in China, the margin rate will also increase to cover the risk. This is because there will be new price from a foreign market during this long holiday. All the Chinese futures markets have the price limit rule. All the exchanges will not set the margin rate smaller than price limit rate, so that the daily price move will not exceed margin rate. This is the safest way for the CCP, but may cause a slow reaction in the market to a sudden large price movement.

Here is an example to calculate the balance using Mark to Market and margin rule. It is still based on product cu, and the margin rate is assumed to be 7%. In this example, we do not consider any kinds of fee.

Day	In/Out	Trade	Settlement Price	Position	Margin	Position Profit	Close Profit	Balance	Available
1	100,000			0	0	0	0	100,000	100,000
			Cash in 100,000, hence balance is 100,000						
2		Buy 2 @50,000	50,100	2	35,070	1,000	0	101,000	65,930
		Balance is changed by sum of two profits, available is balance − margin							
3		Buy 1 @50,300	50,700	3	53,235	8,000	0	109,000	55,765
The Position Profit included 2 lots from 50,100 to 50,700, and 1 lot from 50,300 to 50,700									
4			50,500	3	53,025	−3,000	0	106,000	52,975
5		Sell 1 @50,000	49,900	2	34,930	−6,000	−2,500	97,500	62,570
The Position Profit is 2 lots from 50,500 to 49,900, and Close Profit is 1 lot from 50,500 to 50,000									
6			49,700	2	34,790	−2,000	0	95,500	60,710
7		Sell 2 @49,500		0		0	−2,000	93,500	93,500

Notice that by using Mark to Market rule, though the client may make or lose money from their positions, these profits and losses are realized immediately and these positions are at a value of zero. Therefore, futures positions are not assets. They should not be booked in accounting sheets, nor can they be used as a mortgage, like stocks.

It is important for the CCP to find a good settlement price that comes from the market that represents a reasonable price of the instrument at the end of trading hour, and cannot easily be manipulated. Most Chinese futures exchanges use the value-added average of the trade price for the entire trading day as the settlement price. The weight is the trade volume. It is obviously not suitable to represent the end of day price, especially when there is large price movement near the end of trading hour. The advantage of the settlement price is that it is quite hard to manipulate. CFFEX has a different settlement price, which is a valued-added average of trade price in last 5 minutes. It is a better price for the end of trading hour. There are several other special rules to deal with situations that have no trade made for an instrument in the whole trading day.

In the trading systems of the Chinese futures exchanges, calculation based on the Mark to Market and margin rules will be processed all the time. Because no settlement price is determined in the trading hours, the trading system will use the last price instead. Furthermore, on any open order received, the trading system will freeze the maximum potential margin in advance. This leads to a lot of calculation on part of the trading system. We may notice that if a client opened at a different price several times, and then if he closed part of his positions, we may not be aware of which position he closed, so it is unclear which part is position profit, and which part is close profit. Only the sum of the position profit and close profit is fixed. There may be three potential ways to deal with this problem: using first in first out order, using last in first out order, or using the average open price. Different exchanges may use different ways to determine which part of the sum belongs to either close profit or position profit. Though the difference may not have influence on balance, there is another rule in China that requires only close profit to be used as margin in future trading; position profit cannot be added if it is positive, but should be added if it is negative. Therefore, it is necessary to divide position profit and close profit.

In the trading systems of Chinese futures members, similar risk calculation will also be done during the pre-trade period. There will be two differences between member side trading systems and exchange side trading systems in this area. The exchange will check the money for member's portion, and members will check the money for a client's portion. The other difference is unique to China. Members normally increase the margin rate defined by the exchange for their clients. For example, if the exchange margin rate is 7%, members may add 2% to the rate for their clients, thus making the margin rate 9%. This difference is to increase members' security. Of course, this mark up may be different for different clients.

To start trading, clients must wire their money to a member's bank account, which is similar to western practice. But clients are required to specify some fixed bank accounts, called the futures dedicated account, when they opened the account. All money transferred to members from this client should go through this dedicated account. And clients can perform their transfer either from the bank system or from a member system. Members will accept only money in most cases; the only other possible collateral is a standard warrant of commodity deliverable in exchanges. It is often discussed to use other kinds of collateral, but nothing else can be used for now.

Trading systems of Chinese futures exchanges will also check the position limit for each order. There are two kinds of position limits: (i) a member position limit and (ii) a client position limit. The member position limit is to control the positions of all the clients of a particular member. The client position limit is to control the positions of one client. Therefore, for each client, even if his position limit is not reached, his order may also be rejected because the member position limit is exceeded. Some clients with very large positions will have to use several different members and split his positions to deal with it. Notice that, it is possible for one client to open multiple trading accounts with different members, but his positions in different members should be added up to correlate with the client position limit. Similar to a money check, when the client submits an open order to the exchange system, it will use the worst possible case, that the order may already be totally filled, to check whether the position limit may be exceeded.

The margin call and force close rule is another important risk control rule in China. Clients may receive a margin call if they do not have enough money to cover margin, or their positions exceed the position limit. Notice that though there is a position limit check in pre-trade validation, it is still possible for a client to exceed the position limit. In some cases, position limit is set to a percentage of the total Open Interest of the market. Clients may not exceed the position limit when they open the position, and later when the Open Interest goes down, and then exceeds the percentage. The client is required to wire in more money, or close some positions when he or she receives a margin call. If he or she does not do so in the first session of the next day, the member should do so instead. If the member does not do so, the exchange will force close. If the client does not have enough money, a force close will select some positions of this client to close, so that the margin required by the remaining positions is lower than his balance. If the position limit has been exceeded, a force close will close part of the position and reduce the clients holding to the position limit requirement. There is another kind of special force close in SHFE, called non-multiple force close. Some products in SHFE have larger lot sizes in delivery than in trade. For example, a lot of copper is 5 tons in trading, but 25 tons in delivery. Therefore, it is required by SHFE that clients should keep their copper positions in a multiple of 5 when the contract is close to expiration. If not, SHFE will force close the remainder of the client's position.

There is another very special risk control rule in China, which is called the **close by** rule. We know that there is price limit in the Chinese futures market, and if the price moves too far, the market will go to one side of the price limit at the end of trading hour. This phenomenon is called a single-side market. In a single-side market case, some clients may win a lot of money, and some others may lose a lot. If the expected price is too far from the price limit, there may be a continuous single-side market in coming days. This may cause a larger loss than expected for many clients, and many defaults can occur in this situation. Exchanges are expected to handle this risk. This is why the close by rule is introduced. According to rules of all the Chinese futures exchanges, if one instrument has a single-side market continuously for 3 days in a row, it will stop trading in the fourth day. And at the end of the fourth day, exchanges will help any investor who placed a close order at the price limit in the end of the third day to close. This prevents a larger loss from affecting the investor too deeply. Some opposite positions are selected, based on a complicated rule which we will not discuss in this book, and the close orders will trade with these selected positions. This means that if you are one of the losers in this large price movement, you can place a close order at the price limit at the end of the third day, which can help you stop further loss. If you are one of the winners of this large price movement, you may expect that some of your positions may be selected to trade with others without your order. You will make some money by this price movement, but it may not be as much as it is expected. This will be a very rare situation. The last time Chinese exchanges did this was during the financial crisis in 2008.

2.6 Transaction Costs

The transaction costs are ignorable in ordinary trading. However, the transaction costs become the major factor in HFT. In Chinese market, there are two parts of costs: (i) the charges from the futures companies and (ii) the charges from the futures exchanges. The charges from the futures companies are negotiable. Generally speaking, the transaction costs in Chinese futures market are a little higher than US market.

Throughout this book, we compute the transaction charge of CFFEX as $0.0025\%\ V$, where V is the value of the asset.

2.7 Differences with Western Market

This section gives a brief list comparing the differences between Chinese futures markets and the Western markets.

	Western Market	Chinese Market
Instrument Symbol	Normally including product symbol, a letter representing month and 1 or 2 digits for year.	Including product symbol, 1 or 2 digits for year, and 2 digits for month.

(Continued)

(*Continued*)

	Western Market	Chinese Market
Position Management	Using net position for clients.	Using gross position for clients, hence, each order should specify open or close. In SHFE, positions are evenly split between today's positions and previous positions, so closed today or closed previously should be specified for the close order.
Pre-trade Position Check	Normally not done in the exchange trading system.	Check position limit for open orders, and check existing position for close orders. Position limit will include the member position limit and client position limit.
Pre-trade Money Check	Normally not done in exchange trading system.	Check the margin required in member's account.
Trade Price Determination	Use first-in price.	Use middle price among buy price, sell price and last price.
Open Interest in Market Data	Disclose the Open Interest for the end of the last day.	Disclose the Open Interest during the trading hour.
Market Data	Use single side to calculate the trade volume, turnover and Open Interest.	In SHFE, DCE and CZCE, both sides are used to calculate trade volume, turnover and Open Interest, which mean that they will be doubled.
Market Data Frequency	Normally provide high-frequency market data based on each order or trade.	Only provide a snapshot of market data, normally every half second.
Close By Rule	No such rule.	In case of three continuous days of a single side market (end at price limit), the exchange will pause trading in the fourth day and arrange clients loss in the large price movement to close at current price.

Chapter 3

Some Basic HFT Strategies

We explain in this chapter a few basic high-frequency trading (HFT) strategies, most of which are seen in various literatures. The further discussion needs some background in probability theory and statistics. Therefore, we can only start the related discussion from Section 5.4.

3.1 General

Based on our understanding of market microstructure, we can define many different types of HFT strategies. There are four types of HFT strategies that are widely used almost in any market, including

- Arbitrage;
- Ticker Tape Trading;
- Market Making;
- Event Driven.

The following sections will describe each type of HFT strategy, including the basic idea or assumption of the strategy, analysis methods and examples. The last section is on several other kinds of HFT strategies. Note that all these strategies were originally traded manually. Because of competition and technical improvement, more and more manual trading strategies become automatic, which makes the strategy faster, more accurate, more stable and easy to predict. This is just the basic idea of any HFT strategy.

3.2 Arbitrage

When used by academics, an arbitrage is a transaction that involves no negative cash at any probabilistic or temporal state and a positive cash flow in at least one state; in simple terms, it is the possibility of a risk-free profit at zero cost. For instance, an arbitrage is present when there is the opportunity to instantaneously buy low and sell high at no transaction costs. The precise mathematical definition of arbitrage is given in Section 9.1.

3.2.1 Definition of Arbitrage

Arbitrage is the most popular strategy for HFT. At the very beginning, arbitrage was referred to as a strategy that takes advantage of price differences between two markets for the same instrument, or a strategy to buy in the market with lower price and sell in the market with higher price at the same time. Such a strategy can be used for different exchanges trading the same instrument. Sometimes, this kind of arbitrage is called deterministic arbitrage. Since there is no same instrument traded in different exchanges in China, there is no possibility to participate in deterministic arbitrage in China.

In practice, another kind of arbitrage is more widely used, which is called "statistical arbitrage". In our opinion, **a statistical arbitrage strategy is to repeatedly trade a basket of assets according to the same algorithm and get an accumulated profit with statistically stable positive rate**. We have no restriction of the minimum number of assets in our basket, which is different to the currently popular definition of statistical arbitrage that used to need at least two assets (see Ref. [21] for example). We will show in Chapter 7 that we can get statistical arbitrage just from repeatedly trading one asset by a same algorithm. However, in order to understand our algorithms, some basic knowledge of modern probability theory and statistics is necessary. That is why we are only able to show those examples of statistical arbitrage after the main part of Chapter 5.

Here is a simple example of statistical arbitrage. Let us consider the gold price in US market (CME) and Chinese market (SHFE). We can assume that these two prices are correlated highly. That is, it is impossible that the US price increases significantly while Chinese price is stable or even goes down. However, in the short term, there may be some differences between these prices. If we find that the US price goes up by US$20 per ounce, and the CNY/USD exchange rate is 6.2, it means that China's prices should go up for:

$$20\frac{\text{USD}}{\text{Ounce}} \times 6.2\frac{\text{CNY}}{\text{USD}} \times 0.0322\frac{\text{Ounce}}{\text{Gram}} = 3.99\frac{\text{CNY}}{\text{Gram}}.$$

If we found that in the same time, the Chinese price only goes up by 2 CNY/gram, then it means that the two price movements do not match. We do not know which price is more correct. It may be that the US price is too high; maybe the Chinese price is too low. In such a situation, we can buy gold in China, and sell gold in the US at the same time. If in the future, the Chinese price goes up to cover the 1.99 CNY/gram gap, we can make a profit in the Chinese market. If the US price goes down, we can make a profit in the US market. When we find that the two prices are close to equal, we can close both positions simultaneously. This is the typical process of statistical arbitrage.

Notice that there are several kinds of risks in this strategy. The first risk is that there is a factor, the exchange rate, which may be changed from time to time. If just after our opening operation, the exchange rate suddenly goes down, the US price and Chinese price may match now, and we will make no money on this trade, or even lose some. However, if we keep the position only for a very short time, the exchange rate may not move so fast. Since the exchange rate has a very large influence in the economy, the probability of moving too fast is not high. Therefore, it is a risk, but not a large one.

The second risk is that the US price and Chinese price will not match again for a very long time. If so, we will have no opportunity to close positions. This is a critical risk. But we know that gold is a high-value precious metal, and the logistic cost is very cheap compared to its own worth. If the US price and Chinese price never matches, one can spot trade and easily make money on it. So banks on both sides are not willing to see this kind of situation. Therefore, they will try to keep matching the prices, though it need not be so accurate. Therefore, it is a slightly higher risk than the previous one.

The third risk is that the contract sizes of the two products are different. A lot of US gold contracts are 100 ounces, which is 3,110 grams. And a lot of Chinese gold contracts are 1,000 grams. If we buy 3 lots of Chinese gold, and sell 1 lot of US gold, there will be 110 grams of exposure in this series of trades. If the future prices go up, though the 3,000 grams part of the trade is safe, the 110 grams part will bring a loss. The loss may be larger than the profit of 3,000 grams from the arbitrage. But if we do this kind of trade many times, this 110 grams exposure will be present every time. However, in statistics, half of the times this exposure occurs make the trader money, and the other times it loses. Since the average should be close to zero, if we trade enough times, we can counterbalance this exposure. So it is a larger risk, which can be mitigated by more trading.

The fourth risk is that we need to do the trading simultaneously in both exchanges. It is possible that we open with our desired price in one exchange, but cannot do so in the other. We call this a miss. After a miss, we can handle this in two ways: close the position we have made immediately, or open at a less than optimum price in the other exchange. In the first method, we may lose the bid-ask spread and commissions. In the second method, we may open at a disadvantageous price, which may not make any money in the future. Neither of them is a good result, but we cannot find a better way. The only thing we can do is to minimize the miss rate, which will depend on better trading processes and IT methods. This is a very significant risk, which may result in a total failure in statistical arbitrage.

According to the above analysis, we can see that statistical arbitrage can make money in a statistical way. We cannot ensure that each time we trade we will make money. Nevertheless, if we repeat it enough times, the average result will be a positive profit. This is why it is called statistical arbitrage. We will understand statistical arbitrage better after the discussions in Chapters 5 and 7.

3.2.2 Different Types of Arbitrage

A **spread order** is a combination of individual orders (called **legs**) that work together to create a single trading strategy. Any combination of instruments, which have a stable relationship based on prices, can be used for arbitrage. In general, there are three types of arbitrages based on the relationship of instruments in the futures market: calendar spread, cross-market arbitrage and cross-product arbitrage.

Calendar spread requires one to trade on diffcrent delivery months with the same product. It is based on the assumption that the expected futures prices for a different delivery time will share the same movement. This is true in most cases. For example, index futures are normally quite stable for the calendar spread. Below is a graph of the price difference of IF1303 and IF1302 on February 1, 2013 based on high-frequency market data. There are three lines, one line represents the difference of last price and the other two are the difference of bid and ask price, the calculation method will be discussed later.

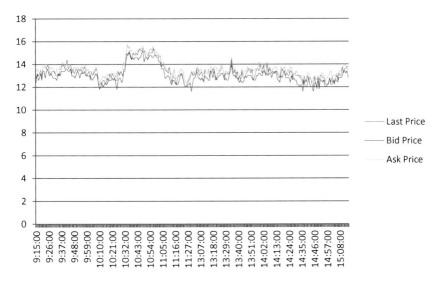

We can see that the range of the price difference is very narrow, but does not stay the same throughout the whole day. Therefore, this pair can be chosen for generating a good calendar spread strategy. It means that its range is not large, so we can predict the price difference in some way. Since it still will change some, we can have some opportunities to trade. We will come back to the details of this issue in Chapter 5.

There is a special kind of calendar spread, called **spot-futures spread**. This type of spread trades spot market versus futures market. For example, we can trade some CSI 300 related spot market instruments against IF futures, such as some CSI 300 ETFs. But as we mentioned before, the spot market uses the $T+1$ rule, which means if you buy ETF today, you can only sell it tomorrow. Hence, it is not considered high-frequency trading in

China. Recently, there have been several other ways to trade ETF similar to the T + 0 rule, but the high cost is still a problem.

Cross-market arbitrage refers to the technique in which one finds the same or similar products in different markets, and trades between them. The previous example of gold trading is a cross-market arbitrage. Here, we also give a price graph showcasing this arbitrage. We still use the market data from February 1, 2013, and select au1306 from the SHFE and GCJ13 from CME. We calculated the price difference of au1306 and GCJ13 and divided by 5. The following is the result.

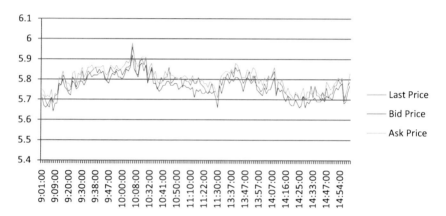

We can see that this graph is very similar to the graph of the IF calendar spread, demonstrating a stable price with a small price movement range. This is a typical arbitrage market.

Cross-product arbitrage refers to using the internal relationship between several products, and trading among them. Cross-product arbitrage is widely used in the securities market in the US and Europe. It assumes that they should have a similar performance for similar stocks. This might occur because both companies are working in same industry. One may also assume that stocks for corporations in one supply chain will have the opposite performance if the final market has no change. This kind of arbitrage is called alpha arbitrage. In the futures market, cross-product arbitrage is often used for products in one step of manufacturing. For example, we know that most parts of soybeans are crushed into soybean oil, and the remaining part is soybean meal. All these three products are traded in the Dalian Commodity Exchange (DCE). Below, we have a graph displaying such a combination. According to some domain knowledge of the industry, about

5 units of soybean can be crushed to 1 unit of soybean oil, and the remaining are 4 units of soybean meals. We use the price of soybean oil (y1305), plus the price of soybean meal (m1305) multiplied by 4, and subtract the price of soybean (a1305) multiplied by 5.

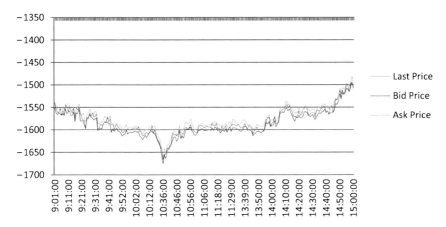

We can see that the price range is much larger than the previous two examples. It means that the cross-product arbitrage is not as stable, a bit more statistical. This forces traders to rely more on mathematical modeling than the previous two types which rely more on IT competition.

3.2.3 Process of Arbitrage

Here, we will give some detail on how to do real trading using arbitrage. There are always many ways to do so, and the method below is just one of them. But the key ideas are same.

Because arbitrage usually traded on several different instruments, the first thing to do is to define a **pseudo instrument**, which is a combination of instruments. Then, we calculate the market data of this pseudo instrument. Let us return to the calendar spread example for IF. Because we hope to trade on the difference of IF1303 and IF1302, we use this definition of the combination. Suppose that the data below is the current market data of these two instruments.

Instrument ID	Last Price	Bid Price	Bid Volume	Ask Price	Ask Volume
IF1303	2,699.6	2,699.4	2	2,699.8	7
IF1302	2,686.6	2,686.6	5	2,686.8	20

Now we should consider the suitable market data for the combination. For the last price, it should be the trade price of the last trade, and for the combination, it should be the difference of the last trade between these two instruments. Therefore, it should be $2{,}699.6 - 2{,}686.6 = 13$.

Now we should consider the bid price and ask price. If we want to buy a lot of the combination, we should buy 1 lot of IF1303 and sell 1 lot of IF1302. To sell 1 lot of combination means to sell 1 lot of IF1303 and buy 1 lot of IF1302. The bid price of the combination should be the price of selling 1 lot of combination using the counterparty price, which means to sell 1 lot of IF1303 and to buy 1 lot of IF1302 both using the counterparty prices, and these prices should be the bid price of IF1303 and the ask price of IF1302. Therefore, the bid price of the combination should be the difference between them, that is: $2{,}699.4 - 2{,}686.8 = 12.6$. In the same way, the ask price of the combination should be $2{,}699.8 - 2{,}686.6 = 13.2$.

Then we should calculate the bid volume and ask volume. The bid volume should be the maximum volume we can sell on the counterparty price at that time. Consider the combination: How many lots can we sell at 12.6? It should be the minimal of bid volume of IF1303 and ask volume of IF1302, that is $\min\{2, 20\} = 2$. And the ask volume should be $\min\{7, 5\} = 5$.

Therefore, the current market data of combination should be as follows:

Last Price	Bid Price	Bid Volume	Ask Price	Ask Volume
13	12.6	2	13.2	5

Based on this calculation, we have a new market data for the combination: all the fields have a similar meaning for a single instrument. Then, we can do analysis and trading based on this market data, just as if it were a normal single instrument.

Now let us discuss the more general situation. There are two more things we should consider. The first is that we may have multiple lots for some legs. Because we are going to find some combination with a stable value, multiple lots are important for instruments with different sizes. We call it as **leg multiplier**. The second is when we calculate the price, there should be a multiplier used for the price of each instrument. We can see this in the example of gold arbitrage above. This multiplier is based on instrument size and exchange rate. We call this multiplier the market data multiplier. Therefore, to define a general combination with n legs, we shall use the following parameters for each leg i,

- I_i, the instrument of this leg,
- D_i, the direction of this leg, 1 for buy, -1 for sell,
- L_i, the leg multiplier of this leg, which must be a positive integer,
- M_i, the market data multiplier of this leg, which can be any real number.

We should notice that the direction of each leg only indicates the real direction when we buy this combination. If we change all the directions to the opposite directions, there is no real difference. Therefore, we can just assume that D_1 is buying, and adjust all the other leg directions if necessary.

Below we will give pseudo codes to calculate the market data of the combination which we will name as C.

```
C.LastPrice = 0
C.BidPrice = 0
C.AskPrice = 0
C.BidVolume = maximum integer
C.AskVolume = maximum integer
for i = 1 to n
{
        if D_i = 1 then
        {
                C.LastPrice = C.LastPrice + I_i.LastPrice × M_i
                C.BidPrice = C.BidPrice + I_i.BidPrice × M_i
                C.AskPrice = C.AskPrice + I_i.AskPrice × M_i
                C.BidVolume = min(C.BidVolume, I_i.BidVolume/L_i)
                C.AskVolume = min(C.AskVolume, I_i.AskVolume/L_i)
        }
        else
        {
                C.LastPrice = C.LastPrice − I_i.LastPrice × M_i
                C.BidPrice = C.BidPrice − I_i.AskPrice × M_i
```

$$\text{C.AskPrice} = \text{C.AskPrice} - I_i.\text{BidPrice} \times M_i$$
$$\text{C.BidVolume} = \min(\text{C.BidVolume}, I_i.\text{AskVolume}/L_i)$$
$$\text{C.AskVolume} = \min(\text{C.AskVolume}, I_i.\text{BidVolume}/L_i)$$

    }
}.

Using this code, we can calculate the market data for any combination. However, some of the definitions of combinations may be not reasonable. A good definition should ensure that the real profit of trading should be equal to what we can calculate in the combination market data. Therefore, the definition must have some restriction. Here, we use MP_i for the multiplier of instrument I_i, and assume that there is a standard currency, and E_i is the exchange rate of the currency of I_i to this standard currency. We have made an open of combination with prices OP_i for one lot, and a close with prices CL_i for I_i for one lot. Thus, the real profit of the trading is,

$$\text{real profit} = \sum_{i=1}^{n} D_i \times (CL_i - OP_i) \times L_i \times MP_i \times E_i.$$

And if we consider the combination, the open and close price should be

$$\text{open price} = \sum_{i=1}^{n} D_i \times M_i \times OP_i \text{ and close price} = \sum_{i=1}^{n} D_i \times M_i \times CL_i.$$

Hence, the profit calculated based on the combination price should be (here we use MPc as a contract multiplier of combination)

$$\text{combination profit} = \sum_{i=1}^{n} D_i \times M_i \times (CL_i - OP_i) \times MPc.$$

These two profits should be the same in any cases regardless of the change in CL_i and OP_i. Therefore, the coefficient of $(CL_i - OP_i)$ should correspond to each other in both equations. That is,

$$M_i \times MPc = L_i \times MP_i \times E_i \quad \text{for any } 1 \leq i \leq n.$$

Normally, we will use leg 1 is a special leg, whose direction is always buy, leg multiplier is 1, and the standard currency is just the currency of the first leg, hence, $E_i = 1$. And we may define the contract multiplier of combination is the same as the first leg, which means $MP_c = MP_1$. Then we know that the market data multiplier for each leg i should be,

$$M_i = (L_i \times MP_i \times E_i)/MP_1.$$

And specifically, M_1 must be 1.

Using this formula, we can define a combination of any structure, and ensure the market data of this combination is reasonable market data, we can use it to calculate the real profit. The key problem of finding an arbitrage opportunity becomes the problem of finding a proper combination, whose price is stable for a long period, and has some movement in a short period.

There are mainly two kinds of known ways to find a suitable arbitrage combination. The first is using some knowledge of economics or industry to find a potential combination, like the calendar spread, cross-market and cross-product arbitrage we have mentioned. The other way is just to hunt through the numerous market data and find a stable combination, and then try to make an explanation of it. The second technique is now a typical one for big data. Different traders may have knowledge from different fields, and may focus on a special area of arbitrage. This will lead to a main difference between traders. From Chapter 5, we will consider a technique which combines the above two techniques and needs more knowledge of probability theory and statistics.

After finding a combination, we should make a decision about the buy or sell price for this combination. The most popular way to decide is to utilize some kind of average history of the prices of the combination market data. We can base our decision on the average of the last prices, or the counterparty price in the different market. We can use a short period market data to calculate, or use a longer period one. A short period may be helpful for not relying so much on a stable arbitrage, such as cross-product combination, and a long period can be used for more stable arbitrage, such as calendar spread. For the beginning of a trading day, we may use some prices near the last close. And there are still several average methods we can choose to use: arithmetic average, value-added average or moving average. All these methods may be effective. The best method can be found only through deep analysis on real market data.

The final problem we should discuss for arbitrage is the trading processes. The core target of a good trading process is to minimize the missing rate, or minimize the loss caused by a miss. Another target is to increase the possibility to make trades. There are two kinds of trading processes, removal and passive.

Removal means to always take the counterparty price for all the legs. One tries to buy on ask price, and sell at bid price. It is similar to using combination market data. The problem is that the real orders are based on each leg; we cannot ensure that these legs are all traded simultaneously. If some legs are traded and some are not, normally we may use some deep price to trade for these non-traded legs, though we know that it may cause a miss, it is still better than keeping unbalanced positions for a long time. For some legs, their order book is thick, which means there are many orders in their order book continuously. We will not lose a lot if we have a miss on them. But legs with thin order books will be very dangerous. There may be a large gap behind the best bid or ask price, and a miss will cause serious loss. Therefore, one of the most popular practices in the area is to find one leg with the thinnest order book, and try to make a trade in this leg first. If the trade is made, then we shall try to trade on all other legs with thicker order books. In this case, we will not have a very large miss. If this leg cannot be traded, we can just cancel the order and do nothing with the other legs. We will just lose this opportunity to trade with anything, without any miss at all. This will be a safer way. Normally, we can predict the thickness of the order book by studying the market, and in most cases, the difference is very obvious. In some cases, we can detect the thickness in trading time, and make the decision on the first trading leg based on it.

Passive trading is to place orders at the desired price in order book, and wait for counterparties to trade with them. While participating in arbitrage, we can only do passive trading in one leg, and after trading, remove all other legs. We can do so at any leg of the combination, but similar to the consideration of removal, we would better be passive in the leg with thinnest order book. There are many kinds of refinement for passive trading. Some are focused on getting a better price, some on placing multi-layer orders to minimize the time gap between order cancelling, and some on getting better time priority for its price level. Traders often give new ideas for these kinds of processes based on the understanding of the market microstructure.

Both removal and passive trading are important processes for arbitrage. Removal trading is simple and safe compared with passive trading.

But passive trading may get a better price than removal, the difference is the bid-ask spread of the market data of the leg performing passive trading. And passive trading will have much more trading opportunities. But since you always put some orders in the exchange, any system problem may cause serious result. And in the Chinese futures market, you need to freeze the money while placing orders; therefore, you may need more money to do passive trading.

3.3 Ticker Tape Trading

Ticker Tape Trading is another important type of HFT strategy. The name comes from the ticker tape, which is the earliest form of digital electronic communications medium: transmitting stock price information over telegraph lines. Later, it became a scrolling electronic ticker screen on brokerage walls and on financial television networks. Ticker tape trading is to trade based only on the information on a ticker tape, that is, just market data.

Previously, there were many manual traders who were doing this job. They were trained to read the market, and base their trades on the market data during the trading hours to find any trading opportunity. They would use some specially designed trading GUI, which would help them trade quickly by pressing fewer keys to send an order. This is called click trading. Later, more sophisticated programs were designed to replace the position of these traders, which led to ticker tape trading.

There is an assumption behind ticker tape trading, that we can make short-term predictions of the price by recognizing some pattern in the market data. There are many such assertions. Here are some examples.

- A large volume of orders in the order book indicates that the future price will not easily go through the order price, and the price will probably go back if this order price is touched. Hence we should identify how a large volume is required.
- A large trading volume indicates price will reverse soon. Then, we should find out how large of a volume is required.
- Instrument A is always a leading signal of instrument B. Therefore, we should find out any price movement of A which has not occurred in B, and predict B will have similar movements.

We are not going to say that those assertions are correct, or if they are correct in some specified market. This is shown to demonstrate what kind of patterns ticker tape trading strategies use. They all only rely on the market data and recent history of market data of the instrument we are trading, or some related instruments. All this information can be found on a ticker tape.

If there is always the same group of investors in one market, their trading activity may be different, but each investor's behavior should be stable. As the market performance is the accumulation of all these behaviors, we can predicate that it may have some kind of internal laws. This is the theoretical foundation to ticker tape trading. As there are different investors in different markets, the internal laws may be different. This is why we should check these assertions for each market independently. And in reality, there may be some investors that leave the market, and some that join every day, which may be a small percentage of the investors. But in the long term, these changes may be quite large. Therefore, the internal laws may have slight changes every day, and accumulate large changes after many days. This indicates that we can use the recent historical market data to check these assertions for the near future. We should often do this kind of work, even every day, to keep pace with the changes in the market. Most of funds that are participating in ticker tape trading will do daily validations of all these assertions after the trading hour. The results will guide the trades in the next day, including which signals can be used, and which parameters should be set.

There are mainly two kinds of analysis methods for ticker tape trading. They are assumption with validation, and mining with interpretation.

For the method of analysis called **assumption with validation**, the trader will try to find a lot of assumptions based on his domain knowledge like the examples above, and use many different ways to fit all the related parameters for each market. The first problem in this method is finding a good way to describe the assertions that are not so formalized. For example, how to describe a quick price move? We should find a mathematical definition for it that can describe the speed of the movement, and also eliminate extraneous information in the market data, such as an error committed by other traders that would cause a sudden movement. A lot of mathematical tools can be used to achieve this. The other problem in this method is to determine the effectiveness of the signals found. Some researchers are willing to use some kind of simulation trading system, and some just check the market data after the signal. It is hard to say which way is better.

For the method of analysis called **mining with interpretation**, researchers will do data mining on a large amount of market data, and hunt for any potential pattern. After finding such a pattern, researchers may try to give an interpretation of this pattern and hope to confirm that it is stable. Normally, all the related historical market data should be marked as a buying or selling point according to the real price movement, and many different kinds of mathematical and IT tools can be used to find these laws. Based on the linear correlation assumption, a correlation analysis or Granger causality test can be used for this purpose. Various machine learning algorithms can also be used on a non-linear situation. Some funds have a huge number of computers that do these complicated calculations every day. There is a different opinion on whether or not an interpretation is required. For some long-term predictions it seems much more important to do so. For a short-term area, some funds insist that they find an interpretation so as to feel comfortable before trading; other funds believe that there will be little risk to keep positions for a short time. Therefore, an interpretation is not necessary. It is not clear which method is better; making money is the final benchmark.

Because there may be many different assertions or mining results, each one will give its buying or selling signal independently. It is important to merge these signals and produce the final signal for trading. It is not suggested to just simply have a union of all these signals, as there are more important features behind these signals. Here are some:

- Difference in performance for each type of signals. Here, performance is based mainly on how accurate these signals are.
- Differences in the frequency of each type of signals. For example, an assertion with 100 signals per day and another with 2 are totally different. We hope to find assertions with more signals. And for some not frequent but accurate signals, we may need to combine them with some other not so accurate but very frequent signals.
- The relationship between different types of signals. As some signals are very similar, we should not regard such kinds of double signals as an enhancement of one signal.

- How to deal with signals that contradict others? Shall we accept the stronger signal, or just ignore all these signals to be safe?
- Shall we use different signals for opening and closing? Maybe, we can use some strong signals for opening, and some weak ones for closing.

There is no standard way to deal with all these problems, and each fund may have their own solutions for these.

Based on the merging of all valid signals, we have the final trading signals for ticker tape trading. In most cases, a trader may use removal trading to trade based on these signals. Some signals will give the predicted price and passive trading may be used in such a situation. We will come back to this topic again in the second half of this book.

3.4 Market Making

Market making is to provide two side orders for one instrument in most trading hours, in order to provide liquidity to the market. In most of the market, there are special rules to encourage a market maker. Normally, market makers are required to provide two side orders with no more than a few specified ticks on the required instruments for some percentage of the trading hour. There also may be some requirements on volume of the orders provided, and sometimes for the depth of the orders are also required. The key benefit of becoming a market maker is reducing the trading cost. For some exchanges, there may be special order types or special market data available for a market maker. But the Chinese futures market is an exception, there is no such kind of market maker encouragement plan there.

There is a lot of work on the theoretical model of market making. For example, the inventory model tries to find a suitable price based on your current positions. It should be based on the assumptions to minimize the impact from the other worlds to this market, which is probably not true. Another widely discussed model is the information model. This model assumes that there are two kinds of investors in the market: those with information and those without information. Investors with information, we can assume, will always catch any large movement of the market. The activities of investors without information are somewhat like random trading. A market maker will make money if he or she trades with investors without information in large numbers, but always lose money when trading with investors with information. Try to find out which part of the market data comes from investors with information: this becomes the key problem of this model. A lot of mathematical tools can also be found to filter random noise from the market data. All these theoretical models can be found in many related books. We will focus on some models based on HFT strategies below.

The most popular market making method is to use arbitrage to do so. It is just like the passive arbitrage described above. Both cross-market and cross-product arbitrage can be used for market making of the main calendar month of futures contracts, and the calendar spread method can be used for the other calendar months. All these methods are based on some reference instrument, which is believed to lead in the instrument market making. There should be some amendments in the arbitrage to fulfill all the requirements for a market maker. Because the market maker should always give a correct price to guide the market, it is important for a market maker to identify any incorrect miscellaneous price in the reference instrument. After identifying these errors, a market maker may use other ways to calculate a correct price, or just give up market making for this period. Some market makers may only use some reference instruments to determine the price without real trading on these reference instruments. This gives them more flexibility to use multiple references. But there will be much more risk without a balanced positions via arbitrage. This way is often used for some instruments that already have good liquidity. As you may not keep positions for long time, the profit gain from the bid-ask spread will be a significant part that can cover the risks.

Another popular market making method is using a signal for ticker tape trading that may give a guide for price in the near future. Normally, a good ticker tape trading strategy only needs to catch part of the price movements, and may ignore others. The accuracy probability is the key benchmark of ticker tape trading strategy. But if we are going to use it for market making, a higher requirement for the probability distribution of each situation will be very important. Data mining may be used to achieve this. By using some consolidation works, we should mark several kinds of market data patterns and give a stable historical price movement distribution. And based on such a distribution, and the expecting profit, we can find the proper buying and selling price in the current situation, and then refine it with the obligation of the market maker.

In the current China futures market, there is no encouragement plan for market makers. Because there are many individual investors in China, there is very good liquidity for most of the main calendar month contracts, which have no need for a market maker. But there are some exceptions, such as the market for steel wire rob and fuel oil. But most of the non-main calendar month contracts have worse liquidity. It is often discussed in China that some encouragement plan for these contracts may be required. This has not yet had an effect. There are many small futures-like exchanges in China who have very good encouragement plans for a market maker.

3.5 Event Driven

An event-driven strategy is quite different to compare with the other HFT strategies. That is a manual trading method formed by the following steps: reads related news quickly, analyzes its impact on the market, and trades before the market is influenced by this news. For example, if we get news about an earthquake in Chile, we may expect that the copper price go up, as Chile has the largest copper mine in the world; and if we get a report that Chinese economic growth is slowing, all the prices on materials may go down, since China is a key consumer of these materials. In these cases, traders may open positions according to the news immediately, and close them after the price becomes stable.

The key for event-driven strategies is to gather all related information quickly and analyze the impact accurately. Nowadays, computers are used to replace manual work for this. Therefore, all event-driven strategy systems will include two parts, a spider and an analyst.

A **strategy spider** is used to scan the Internet quickly and filter all the related information. The requirement of spiders for event-driven strategies is different from those of search engines. It is considered a good result for a search engine to scan the whole Internet in a half hour, while this delay is not acceptable for HFT strategies. However, it is not necessary for strategy spiders to keep scanning the whole Internet. Normally, strategy spiders will focus on the following web sites:

- Official sites for some economic reports;
- Official sites for government announcement;

- Financial reports of list companies;
- Important news sites;
- Some quick news sources, such as twitter.

It may be a problem if we choose these quick news sources, because we cannot ensure the accuracy of the information on these sites. It is even a problem for humans to clarify them, so we cannot expect software to totally solve it. We may require multiple sources for confirmation of this kind of news, but it will waste valuable time.

For all the information retrieved by a spider, we should filter them to get the real important things. To do so, we must first let the system understand the meaning of the unstructured information, various local dialects understanding technologies may be used to do so. And some patterns will be defined to select useful messages. Such patterns may include economic data patterns, disaster patterns and so on.

After consolidating all the related information, a **strategy analyst** should be used to make real predictions. Some analysts use domain knowledge, and manually set up. For example, take the influence of a disaster. It is not easy to accurately use this knowledge. There might be an obvious consequence, like the Chile earthquake, but how about the Japan earthquake? Japan is too large and complicated, and the real influence of the earthquake is not easy to say. The last earthquake and tsunami in Japan in 2011 caused a dropdown in rubber. Why? It is because of the automobile manufacturing industry. And traditionally, we may think that a flood in Thailand will cause rice price to go up. However, the flood in 2011 also halted the work of many large industry parks. Therefore, even manual analysis is not simple. Some other analysts are using machine learning technology to do everything automatically. The key problem is to find enough sample data. There are not enough disasters for this method to work. The reports of economic data are periodical and sufficient. And financial reports of listed companies are a better sample to use. It is always complicated work to do local dialect understanding, and it may take a long time as well. Another kind of method is to perform an extraction of featured vocabularies and data. It is used on many official reports and analysis. Because these reports may have some standard templates, we may understand the meaning by only retrieving some keywords and data.

Though event-driven strategies are widely used by the funds, the risk is quite high. Since it is so complicated to understand local dialects and analyze their influence, a lot of mistakes are made in those funds. Even for feature extraction, the template of reports may have suddenly changed. The performance of event-driven strategies is not stable according to historical statistics, though a lot of profit may be made in a few cases.

Notice that event-driven strategies are not really considered high frequency, and some researchers may not include it in HFT strategies. But it does use a lot of advanced technology which is similar to the other HFT strategies, which is why we still include it here.

3.6 Other Basic Strategies

HFT strategies are an innovation area, and many researchers are working hard to find many different strategies, or apply existing HFT strategies in a new way to a different field. It is hard to list all these strategies, but here are some of them.

A traditional area in large order execution is now being revolutionized by change in HFT. Large order execution requires one to buy or sell a large number of specified instruments at a required price. The traditional way to do so is by trying to place an iceberg order, or simulate an iceberg order, which will most likely have little influence on the market suddenly, so as to avoid a bad price. Because some HFT strategies can have short-term predications of price, these strategies are being used in large order execution now. If the price is expected to get better, no order or few order volumes will be placed in the market. If price is getting worse, a quick order is inserted. Such kinds of arrangement may make for a better execution price. The business of large order execution is moving from large investment banks to funds with these kinds of strategies.

Some strategies are focused on the influence of orders during auction time. Based on our understanding of the auction process, systems can calculate the possible auction price change for a potential order insert. Thus, if some traders are expecting a current auction price execution, they can calculate the maximum volume they can put in to keep this auction price. And they may consider how to get a higher priority to get an execution.

Auction is a complicated process, and there is a lot of research on how to take advantage of this process.

In China, there are several pauses in trading hours every day. When the next trading time comes, some pending orders placed in the last session may have the incorrect price. Some traders are focused on trading with such kinds of incorrect price. The key is to get the first-order insert in place after resuming trading. Because time is not very accurate, it is hard to know when it is a suitable time. Some technologies have been developed to get a better chance to achieve this.

The special rules in China, such as the close by rule may be used for strategy design. If it is probably going to have a continuous three single-side market (see the table in Section 2.7), one may keep positions in both directions. Since the close by rule requires the losing direction to be closed, the wining direction may be kept. This is a simple way to make profit according to the close by rule. Some other structures are also designed to use this rule.

All these strategies are special and not easy to put in any HFT strategy types. Researchers will not be bounded to any restriction.

Chapter 4

IT System

The IT system plays a core role in high-frequency trading (HFT). All the trading, management and research are based on IT systems. In this chapter, we will discuss the details of IT system design.

4.1 Challenges

A typical IT system for HFT includes the following components:

- A trading system to implement all the strategies and handle the trading process.
- A market data system to gather and manage all the required market data.
- A simulation system to do simulating trading and concept proof of trading strategies.
- A research support system to provide all kinds of tools for strategy analysis.
- An internal settlement system to do settlement for internal purposes, such as statistics on the strategy.

Among all these IT components, the trading system is the most important one. The challenge of trading system is the most critical one. The trading system implements all the related functions in competition with other traders. The key benchmarks of a trading system are the following:

- Internal response time;
- Message throughput;
- Adaptive for fast change.

The **internal response time** is the most important benchmark for a trading system, which is the latency between receiving a message and inserting an order triggered by this message. This latency is all about the handling process of the HFT system, and nothing to do with the exchange side system and any outside environment. If a trader has a shorter internal response time than others, he can insert orders earlier than them. Then for all the trading opportunities, he can have a higher priority than others. Therefore, if this trader and his competitor share the same strategy (because a lot of strategies are well-known, especially, the classical arbitrage strategies discussed in Chapter 3), he can take all, and his competitor will take nothing. A shorter response time may lead not only to a difference in profit rate but also a difference in profit or loss. Since the winner takes all, there is an intense competition in internal response time. As a result of competitions, the internal response time for simple strategies keeps decreasing because of competition, from milliseconds down to microsecond nowadays.

The ability to handle large amounts of incoming messages, especially market data, is another important benchmark for a HFT trading system. In some exchanges, the market data that is sent out from exchanges are tremendously huge. In the stock options market, there are hundreds of thousands of options whose market data is changing continuously as the stock price is always changing. If a trading system is going to trade in such a market, it must have a huge capacity to handle all these messages and must be very fast to do so. If a trading system cannot handle all the messages in time, there will be a queue and a delay between receiving messages, which will cause a loss on trading.

Adaptability to change is also a key benchmark for a trading system. The market is changing very fast. The trading systems must be able to follow it very closely. New strategies will appear online continuously, and new surveillance rules must also adapt to the trading system to be relevant. A HFT system may have to release some new versions every week. In many cases, a trader may have no opportunity to test his system before it goes online. A HFT system cannot run independently but must rely on the execution of an exchange system, which increases the difficulty of testing. Many exchanges will not provide simulation environments all the time. Even if exchanges provide simulation environments, it is still quite different between real market data and simulated market data. Therefore, the trader may need to ensure that his system modification is correct and it can go live without suitable testing. Some theoretical correctness proven methods can be used in such a situation, e.g., Hoare Logic.

The design of IT system is always a trade-off between different features. If we want to emphasize the importance of something, then something else may be sacrificed.

Security will usually be sacrificed. Here, the security is based on IT considerations, which includes authentication, authorization and encryption. It seems obvious that the security is a key feature for trading system and for any similar financial system. However, keeping high security means a very high cost for every step, and it will have a large impact on the internal response time. For example, if we define a complicated authorization hierarchy, and check if it is right each time, it may add some microseconds to each response. But we can use other methods to have a similar effect, such as a management guide, operation camera, different operation system accounts and so on. Those methods will have no impact on the internal response time. The encryption is also not necessary inside a trading system, and will have a large impact for each message transfer; even if we use the best encryption chip.

Another aspect to be sacrificed is robustness, which, from the IT point of view, is mainly related to the stand-by system and fails over. A real-time, hot stand-by system costs a lot in each message handling, and a real default due to hardware is seldom nowadays (software problems normally cannot be solved by stand-by systems). Therefore, most trading systems will only have cold stand-by ones, which can only be restarted manually.

4.2 Trading System Design

A typical trading system should be a message-based system, which uses a message system to connect the following components:

- Trading Interface;
- Trading Process Control;
- Risk Control and Surveillance;
- Strategy Implementation;
- Monitoring.

A typical architecture of a trading system is shown as follows:

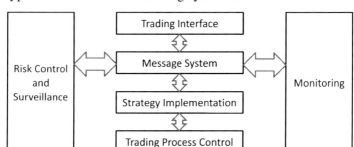

4.2.1 Trading Interface

A **Trading Interface** will connect a trading system to a higher system to get market data, place orders and get notifications. The higher system may be either an exchange trading system or a broker trading system.

Selecting a reliable trading interface with all the functions required from various kinds of options is very important. In general, there are two types of trading interface:

- Direct Market Access (DMA);
- Broker Side System Trading Interface.

The DMA interface is provided by the exchange. If a trader is using DMA, then he can connect to the exchange directly without any control of broker. The broker must trust this trader's system to do all the necessary risk control. The broker can only check it after trading, not before trading.

The broker side system trading interfaces are provided by the broker. A trader's inserted order must be checked by the broker system before it is sent to the exchange. Most brokers will use some third-party broker side trading system. In such a situation, though the broker trading system is run by a third-party IT company, the broker will set all the related risk parameters for each client. Only a few brokers are using home-made trading system.

It is obvious that the DMA is always faster than a broker side system trading interface, regardless how short the broker side system's response time is. However, it will be harder for traders to get DMA. Normally, brokers will have a lot of requirements for the traders to use the DMA, such as capital requirement, minimal trading volume requirement, risk control test and so on. And most exchanges may require some certificates and interface test before one can connect to them. On the other hand, connecting to broker side system trading interface should be totally open for all the clients and will have no other requirements in most cases. Another benefit of the broker side trading system interface is that a trader may trade in a lot of exchanges when he connects on such an interface, to compare with that one can only trade in one exchange by one DMA. If a trader's strategy relies on speed, he must always try to get DMA. Most of HFT strategies should use DMA. Only a few exceptions, e.g., cross-market arbitrage, as a trader may not be able to DMA to all the related exchanges in the same time if these exchanges are far away.

Most trading interfaces will be provided with the **Application Programming Interface (API)** or network protocol. For API, a trader will get some interface description file (such as C/C++ head file) and a library (in most case, dynamic link library). A trader can program based on this description file and link it with the library. In such a case, all functions can be used just by some function calls. For network protocol, a trader may get a description on all the packages and fields of the network protocol, as well as inter-operation between his system and a higher system. The traders should do the network programming by themselves. It will be harder to use network protocol than to use API, and API may hide some details of the real protocol which may be helpful if exchanges have some upgrades. But, because API is very simple to use, there will be no difference among the traders. If someone has the skills to develop a high performance protocol implementation, network protocol will be a better choice.

Most of trading interfaces include two parts: the first is order/trade interface, and the second is a market data interface. The order/trade interface will allow a trader to insert order, cancel order and get the order notifications, trade notifications and so on. The market data interface will provide market data as required. We list in the following space some important features for consideration when one selects a proper trading interface.

For order/trade interface, first of all, we should check whether it is working in query mode or push mode. **Query mode** means that a trader should send a query request each time to find whether or not his order is filled. **Push mode** means that the higher system will send a message immediately after the order is filled. Query mode is a synchronized mode, and it is easy to program, and push mode is an asynchronized mode, which is faster. High-frequency traders should always use a push mode trading interface if possible. The second thing we should consider is how to resume the mechanism after reconnection. Though it is very rare to disconnect, we are not going to abandon all the remaining trading hours. On the reconnection, the trading interface must tell whether or not the last orders in the previous connection were inserted, and what happened to the pending orders during the disconnection. A well-defined trading interface should be well designed for this. The third thing is the detailed function difference between trading interfaces. Some functions, such as inserting order and cancelling order must be included in the trading interface. However, some enhanced functions may not, such as changing order, sending FAK (Fill and Kill) orders, sending market orders. We may simulate some functions if the trading interface does not provide those functions, but they may not be exactly the same. For example, changing an order to reduce order volume will not affect the time priority of your order in most of exchanges. However, if someone wants to simulate it, he can only cancel it and insert a new order, which must have a new time stamp. It is not necessary to ensure that the trading interface has all these enhanced functions, but depending on whether it is important or not for your strategies, you may want to include these functions. The last thing to consider for an order/trade interface is how to get all the supporting data for trading, including instrument information and clear information. The **instrument information** includes all the possible trading instruments, their ticks, multiplies and all related parameters. **Clear information** includes your over-night positions and account information. These data are essential for a trading system during initiation. Some trading interfaces will provide those things directly; others may need a third-party input to do so.

For market data interfaces, we should first check which protocol the interface is using, TCP or UDP? A TCP-based market data interface may give users a continuous market data flow. If there is too much market data which needs to be sent to you, there may be a queue, which causes a delay of all future market data. If it is not acceptable for a trader to base his strategies on market data several seconds ago, a UDP-based market data will have a benefit that some market data will be discarded in congestion.

It will be helpful to recover from such a situation quickly. But it means that the users may lose some market data, and it may affect to the strategies which are sensitive to historical market data. Therefore, a trader needs to select the right market data based on his strategy types. The second thing we should notice is the way in which market data is provided. Some market data is provided by snapshot, and if the user gets a new snapshot, all old information is expired. Some market data is provided by flow, which means a snapshot is provided at first and the change of the market data afterwards. In such cases, we should have some mechanism to provide any missed market data. Otherwise, the error of the market data will be held for long time. A well-defined market data interface should have a way to do so. Some typical ways are providing a query snapshot interface, or periodically resending a snapshot.

After selection of the proper trading interface, we should design our module for it. As we may deal with multiple different trading interfaces, it is suitable to define an internal trading interface, and then map it to the different real trading interface. All other parts of the trading system will only deal with the internal trading interface, which hides any differences of the real trading interfaces. A typical architecture of trading interface module is shown below.

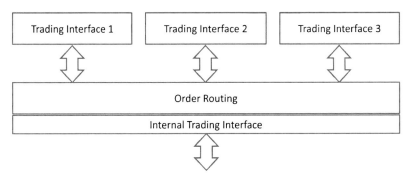

The key problem of designing this module is how to define a suitable internal trading interface. We can define a simple one which may provide only inserting orders and cancel orders, or we can define a complicated one which may add changing orders, FAK orders, market orders and so on. If we are using a complicated design, and the real trading interface also has similar functions, then it is easier to map. However, if the real trading interface is not supported, we may send several different orders to simulate the function. The following are the methods:

- **Changing order**
  Cancel order and then insert a new order with changed parameters.
- **FAK order**
  Insert order and then cancel it immediately after order notification is received.
- **Market order**
  Insert order at price limit, and cancel it if still some volumes are not filled.

We should notice that some other functions may not be simulated, such as **FOK** (Fill or Kill) order and **MV** (Minimal Volume) order. These functions are not suitable to occur in internal trading interface. Even if the functions which can be simulated, we may also have to consider whether or not they should be implemented here. Adding these functions will benefit the real trading interface which provides these functions, but it will also increase the complexity for other real trading interfaces. We should carefully select these extended functions based on strategies.

Another thing we may consider when designing the trading interface module is whether or not it is a good place to plant a simulating trading system at the location. It will allow all the strategies to run in the same way for real trading and simulating trading and also provide a way to test other parts of the system.

4.2.2 Trading Process Control

Trading system is not only used to insert orders, but also used to control the whole life cycle of all orders. A real order will tell exchanges the expectation of traders, but it may not be the best way for strategies to describe their desires. Therefore, another kind of order, we named strategy order, should be defined to ease the implementation of strategies.

The **Strategy Orders** are different from real orders, as strategies may require a series of actions to do to achieve some targets. Therefore, strategy orders may have a lot different parameters, which indicate the expected process. And in reality, each strategy order should map to several real orders.

And further real orders of a strategy may depend on the result of the previous real orders.

There are a few strategy orders which should be defined according to the requirements of strategies. We discuss here three standard strategy orders, which are widely used in HFT. They are the following:

- **Tentative order**
  Try making trade at a desired price and cancel if we cannot make it.
- **Mandatory order**
  Enforced to make required trades in required volumes and achieve better prices if possible.
- **Waiting order**
  Try making trades at desired prices and wait if we cannot make them immediately.

The typical steps for tentative order are shown as the following:

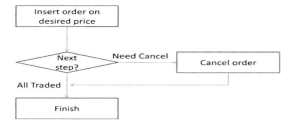

The differences among tentative orders are how to determine the condition to cancel order. The following are some widely used conditions:

- Immediately;
- When time is expired;
- When price is changed in market.

The real condition to cancel order will vary according to different strategies. Therefore, some parameters of tentative orders should indicate which condition to be used. If we are going to cancel an order immediately, we may also use a market order instead of a limit order if the internal trading interface supports the action.

The typical steps for mandatory order are shown as the following:

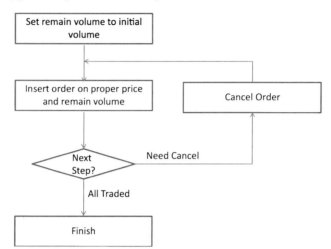

The key problem of a mandatory order is to define the proper price. The following are some possible options:

- The counterparty price;
- The counterparty price plus or minus 1 tick;
- Self side price;
- Self side price plus or minus 1 tick.

During the different times of the loop, we may choose different prices. At first, we may expect a better price, but at a later part, we may look to reach a desired trade volume as soon as possible. Therefore, a reasonable price list for each step of the loop may be as follows (assuming it is a buy order):

1. Bid price
2. Bid price + 1 tick
3. Ask price − 1 tick
4. Ask price
5. Ask price + 1 tick.

The real steps of these prices should be parameterized. Strategies should be left to decide which one is the best.

The other problem of a mandatory order is to decide when to cancel the order, which should be similar to a tentative order.

The typical steps of waiting order are shown below.

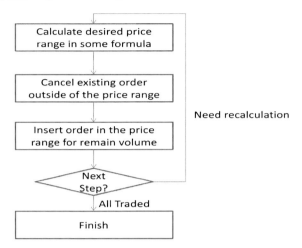

For the waiting order, one may use a lot of orders which will eventually be canceled; hence we may prefer to use a price range instead of a fixed price to reduce order numbers. Larger ranges lead to fewer orders, but also cause inaccurate result and less opportunity to make a trade. Therefore, the range size is a trade-off. Another problem is when to do a recalculation. Normally it will be during a price change in the market, or any change related to the formula.

All these types of standard strategy orders are driven by incoming messages and internal statuses. They should be implemented in the **Finite State Automation (FSA)** model.

All these standard strategy orders are focused on single instrument trading. If some strategies need to trade multi-instruments, a complicated strategy order which is a combination of these standard strategy orders will be necessary. This is especially useful for arbitrage strategies. Below are two of the most useful complicated strategy orders.

- Start a tentative order. If the trade is made, start some mandatory orders.
- Start a waiting order. If the trade is made, start some mandatory orders.

The first one is widely used to do a removal in arbitrage. The tentative order is normally given on the least liquidity instrument, and mandatory orders are for other legs. The second one is used to do passive arbitrage.

The waiting order may be on any leg, and the mandatory orders are for the remaining ones.

By defining these strategy orders, we can control trading process easily. The implementation of strategies can be focused on the signal calculation. Then start some strategy orders with suitable parameters when signals are found and without worry of the procedure of trading.

4.2.3 Risk Control and Surveillance

The importance of risk control and surveillance functions cannot be emphasized enough. As all the trading is automatic, lack of risk control and surveillance means that any error will go much farther than expected. When we have to stop the system, unrecoverable losses may already be made. The risk control and surveillance rules are complicated, and it may take up the majority of internal response time.

There are a lot of risk control and surveillance rules, which are based on instrument, product, exchange or accounts. The following are some popular rules:

- **Position limit**
  Control the upper limit of the position of a specified instrument, or the sum of all positions of instruments for specified product.
- **Single-order limit**
  Control the upper limit of the volume of single order. Sometimes, control the lower limit of the volume of single order, which means that the quantity of your order must be a multiple of it.
- **Money control**
  Control the margin of all positions not to exceed the balance of the account.
- **Illegal price detection**
  Ensure the price is within reasonable range, such as not exceed price limit, or not too far from the current price.
- **Self trading detection**
  Ensure the orders from different strategies will not cause any possibility of trade between them.
- **Order cancellation rate**
  Calculate the order cancellation situation and ensure it does not exceed the limitation of exchange.

- **Flow control**
  Flow control is normally a technical restriction for speeds of input messages. This rule should ensure no real flow control restriction is reached by internal controls.

  As not all the risk rules are applied for each instrument and these rules may take a significant time to deal with, we should use some dynamic rule mechanisms, and only check necessary rules for each instrument.

  Every time we are going to check risk control and surveillance for some specified order, each required rule must be calculated and a restriction must be provided based on it. A typical restriction description generated by rule i should include two integers:

- Order volume limit: L_i;
- Order volume multiplier: M_i.

It means that our order's volume cannot exceed L_i and must be multiple of M_i. As there are several different risk rules, and each rule will give a different restriction, all these restrictions should be combined in the following way:

- Combined order volume limit: $L = \min(L_i)$;
- Combined order volume multiplier: $M = \text{LCM}(M_i)$, where LCM means least common multiple.

And the final order volume should be floor $(L/M) \times M$. Here, floor means the largest integer less than or equal to it. If the final order volume is 0, it means we cannot place an order.

There are three time options that can be used to check these risk rules, which are

- Before strategy calculation;
- After strategy calculation and before order inserting;
- After trading.

The first two options are before trading, which can avoid any excess of these restrictions. The last one is after trading, and it will change some trading according to it. This may have some risks, but it has no impact on internal response time. We may select some rules which are very serious to check before trading, and others to check after trading. For the first two options, we still need to make a selection. We know that in most cases, the strategy calculation will tell you there is no need to trade, in these cases, using the second option will save the checking time as you need not perform any check. But in these cases, because no order was made, there is no need to save time. For the cases where you really need to insert order, these two options are almost same. However, if we consider a system with multiple strategies, each incoming message may trigger several different strategies that need to be calculated. In such cases, the saved time in the second option may be helpful for later strategies. Therefore, if we have to do a pre-trading check, the second option should be a better choice.

Now let us discuss some considerations for implementation of selected rules. First of all, we should notice that there is some basic information which can be used for multiple rules. We should not recalculate this information for each rule, but find a suitable way to share it.

Let us consider the implementation of the position limit rule. In order to know whether his order may exceed the position limit, a trader must have a position keeper to perform a live calculation of the positions for each instrument. This position keeper should listen to all the orders and trade information, so as to accurately calculate. The trader needs to decide whether he should use net position, or use gross position in his position keeper. Most exchanges use net position for traders, but there are some important exceptions, including all the futures exchanges in China, which use gross position. For our internal position keeper, if we use one type, we should simulate the other type for exchanges with the different type. We can see that it is much easier to simulate net position via gross position; hence it is a better choice to use gross position for our internal usage. Another thing we should consider is that we should not only calculate the real position of each instrument, but also notice there are many potential positions. Here is a list of these potential positions, suppose you are the trader:

- **Potential positions by inserted orders**
  They are caused by your previous orders waiting in the order books of exchange. These orders may be filled before you cancel them.
- **Potential positions by inserting orders**
  They are orders inserted by you, but with no order notification received from the exchange yet. The exchange may or may not accept these orders. These orders may be filled before you cancel them.

- **Potential positions by strategy orders**
  Some strategy orders, such as a mandatory order, will ensure you to open or close some positions. Though there may be no orders now, it will automatically insert orders in the near future, so we should still regard it as potential positions.

As there are many kinds of potential positions, we may determine the restriction calculation principle. Shall we use the worst case? In the worst case, if we are going to buy, we should consider all the potential positions with buy orders, and ignore potential positions with sell orders. It indicates the worst situation for buying, all buy orders are filled, and all sell orders are not filled. If in such case, the position limits are still not exceeded, we can ensure that the rules are well checked.

Now let us consider the money control rule. Similar to the position limit rule, we should have an account keeper to calculate the current account information, which must take into account all order and trade information. And the account keeper should also use the position keeper's information to calculate margin. The account keeper should calculate the following information:

- Initial balance;
- Profit/loss;
- Margin for positions;
- Margin for potential positions.

The money control rule will have the same problem as the position limit rule. Shall we use the worst case? If so, all margins for potential positions should be regarded as the same as the margins for positions. Another problem is whether we should use profit/loss in real time? The rule is different in different countries. For example, in China, we should use the loss part of a position profit, but not use the profit part. It is also a problem that the margin calculation may be very complicated. Many clearing houses have a lot of margin deduction rules. If our trading system calculates these complicated rules, we cannot expect a best response time. Therefore, we may give up some rules, and try to find an easy way to calculate it, which should be larger than the real one. It means that we will waste some money because of this, but achieve a better response time.

The self trading detection rule is to avoid any self trading. In order to predict any possible self trading, we should maintain an account order book; including all the live orders of this account. The account order book listens to all order and trade information. If we are going to buy at some price, which is higher or equal to the price of some sell order in our own account order book, we should stop it. Though it is possible for our buy order may trade with other orders with a higher priority in the exchange order book. Because we do not know the priority situation of our own sell order, we cannot take this risk. The key problem with self-trading detection rules is how to find the best data structure to implement the account order book. The operations to account order book includes adding one order, removing one order and getting the highest/lowest price of these orders. It seems that an ordered tree with order based on price is a suitable choice. But actually, a much better choice is using a heap with order on price. It is very easy to implement. Add one order and get the highest/lowest price using heap. To remove one order, we can just mark it to be removed, and only if the top of the heap is marked to be removed do we really remove it and rebalance. Though the complexity level for an ordered tree and a heap are the same, the coefficient of heap is much smaller, and the balance of heap is better. Therefore, heap is the best data structure for it.

We have come to the last rule to be discussed, the flow control rule. Most interfaces have some flow control restrictions with no more than n input messages in k seconds. We have two kinds of attitudes toward flow control restrictions: actively or passively. To actively handle it, we should control our messages to the higher system ourselves, so that the flow control restriction will not be touched. The passive way is just to do nothing and let the higher system queue messages instead of restrictions. If we handle it in active way, we may internally cancel orders above this restriction, and let the strategies or the strategy orders make the decision on whether or not they want to resend the order. In a passive way, we will lose control, and our future orders may be delayed because of the restriction of this second. Therefore, we always suggest handling flow control restrictions actively. However, there is a problem with the phase difference of time. For example, the restriction is 10 messages in 1 second and the start of each second for our system and higher system may have some difference. Consider the following situation:

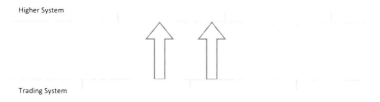

Higher System

Trading System

Each arrow indicates 10 orders. From the view of trading system, the second arrow is in a new second; therefore it is OK to send orders. However, according to the time of upper system, because of the phase difference, it may consider these two arrows to be in the same second, and queue them. We can easily deal with this by reducing the restriction, in this case, use five orders per second. This will ensure that there is not a touch to the restriction. But this solution will cause a large waste of interface. A better way to solve this problem is to keep a queue of the recent n messages' time, and each time we want to send a new message, check the difference of the current time and the time of n messages ago. If it is larger than k seconds, it is OK. Using this algorithm, we can ensure that in each continuous k seconds, we will send no more than n messages. And the waste on interface is minimized.

In some cases, we should use multi-accounts for a single trading system. For example, you are going to do cross-market arbitrage, and the two markets do not share a same clear member. It is also caused by the requirements needed to deal with some surveillance rules. In multi-accounts cases, the risk control and surveillance module should be the best place to manage an account selection algorithm. There are several principles that should be considered for this algorithm, trading rights, trading possibility and balance among accounts.

4.2.4 Strategy Implementation

In a trading system, a strategy should include at least the following three parts:

- Logic to do trade;
- Strategy orders currently running;
- Positions managed.

A strategy implementation should be a message-driven object using subscribe/listener model. It should listen to all related market data, and all related order/trade information.

The key part of strategy implementation is the logic of trade. Here, we will face a question: shall we define a script language (using an interpreter) to describe this logic, or to use a normal computer language (using a compiler) to do so. Below, we list some key considerations for comparing them.

- **Flexibility consideration**
  Using script language makes the strategy easier to develop. And you need not recompile the whole system while performing some minor change with strategies. Therefore, a script language is better.
- **Performance consideration**
  Using a compliable normal computer language to implement strategy will make it run much faster. Therefore, a normal computer language is better.
- **Reliability consideration**
  Here, we may only consider what is needed to not cause the whole system to crash because of mistakes in strategy implementation. Because a script language will have some protections, and a well-designed script language implementation will not be crashed by any script, a script language is better.
- **Team structure consideration**
  As in most of HFT funds, there will be an IT team to maintain the main parts of the trading system, and a lot of *quants* (short for people who perform quantitative analysis) to develop various strategies. It should be better to have a script language for quants which will not impact the codes of IT team.
- **Learning curve consideration**
  The learning curve of script language should be shorter than normal computer language. Therefore, a script language is better.

In summary, except for the performance consideration, a script language is always a better choice. However, the performance is very critical and we cannot ignore it. Therefore, the best choice is to make some kind of balance. Which means one must design a compliable script language that can be used via an interpreter in the research and simulation period. And in the production environment, it can be compiled and kept as fast as a normal computer language.

4.2.5 Monitoring

Though HFT is automatic, it is still necessary for traders to understand what is happening now, and determine if everything is running well. Therefore, a monitoring system is necessary, which should consolidate all the information of trading processes and strategy statuses, and present them to traders in real time. The traders may have some operation on these strategies, at the very least traders should be able to stop them.

We hope to minimize the impact of a monitoring module on the internal response time. Therefore, we may always try to move the work load outside of the trading path. The following are two basic principles to design a monitoring module.

- Do not write log during the process of trading, but write it in the handling of some insignificant message, such as a time notification.
- Handling monitoring communication in independent thread/process.

4.3 Environment

4.3.1 Programming Language

The programming language is the first thing to be considered for trading system environment. There are several obvious options, including

- C/C++;
- Java;
- Assembly language.

Our focus should be finding the fastest programming language. Assembly language is a low-level language, which should be regarded as the fastest one. However, currently the difference of the speed between a high-level language and a low-level language is not obvious, and it is not easy to program a complicate system in a low-level language. Therefore, almost no one will try to develop the trading system using the assembly language, though there may be some critical small part using assembly language.

There is a long history of the debate between C and C++. C is regarded as much closer to low-level languages. Thus, C gets many supporters. The others believe that a well-designed C++ program can run as fast as C, and C++ makes the system clearer and easier understood by humans. Currently, most of HFT developers select C++.

Java is playing a more and more important role in HFT. This is a much younger language and absorbs many new ideas of programming. Java is designed to compile to byte code and an interpreter should be used to run these byte codes. Therefore, it seems as if it is not able to run as fast as C++. But a lot of wise brains have devoted themselves to improve it. Nowadays, a **Just in Time (JIT)** compiler has been introduced to Java, which can translate some important segments of byte code to machine code to improve speed. And some code generation optimization can be done based on the real execution status, which may be better than static optimization in compile time. Some new style HFT designers are using Java, and have got some encouraging results.

4.3.2 Server and Operation System Selection

There are mainly three options for server and operation system selection:

- PC Server/Linux;
- PC Server/Windows;
- Midrange Computer/Various Unix.

Currently, the CPU speed of PC server is faster than midrange computers. PC servers are widely used everywhere, and more popular than the midrange computers. Therefore, all new technology will be used in a PC server very quickly, and mature soon after. Recently, some of the latest technologies are only supported on a PC server. Linux is a better choice than Windows for HFT, as it is simple, open-source and popular server side. Therefore, a PC server/Linux is a popular selection for the HFT system.

Some HFT systems are using optimized Linux. As an open-sourced system, it is possible to refine its performance by changing some low level operation system code. Some others are using some advanced Linux functions to improve response time. Here are some of them.

- **Network protocol stack optimization**
  Some network protocol parameters can be changed, so as to minimize the waiting time in TCP protocol. And some new network protocol

implementations are faster by reducing the time of memory copy and so on.

- **Dedicate CPU core for each thread**
  We can assign a dedicated CPU core for some thread, so that the cache hitting rate can be higher.
- **Use spin lock to replace mutex**
  Spin lock checks continuously until some condition is matched. Use of spin lock will reduce the time of sleep caused by normal locks, such as mutex. It is helpful to keep the control of the CPU by our key trading threads.

Another approach is using some special hardware to improve speed. Here are some examples.

- **Special network adapter**
  Some network adapters are designed to implement many protocols in hardware, which may improve the performance of network implementation. Some other network adapters are using refined network protocol instead of TCP/IP, so that any network between these adapters is especially fast. Some of these new protocols have become standard, for example, InfiniBand, which is playing a more and more important role in trading system.
- **Special chips**
  Some special chips are designed for trading purpose. In theory, we can move all of our logic of trading system to chips. However, the strategy logic is changing fast; hence, it is not suitable to put in a chip, as any modification on chip will take a long time. Nowadays, some risk control and surveillance rules can be implemented in chips, as they will not be changed frequently.

4.3.3 Network Environment

Network environment plays a key role in HFT system. A shortage of network environment may cost milliseconds, not microseconds. We cannot recover from it in any way.

Colocation in exchange is always the best choice for HFT, and it must be the quickest way to access the exchange. Now high-frequency traders are focused on some details of colocation, which may lead the competition to a still higher level.

Many colocation providers will have several different bandwidth options for investors. Some are wider and expensive, and some are narrow but cheaper. It is important to always select the widest one if you are really sensitive to speed, even if your real requirement of bandwidth is not so high. The following is a diagram to show the influence of different bandwidths.

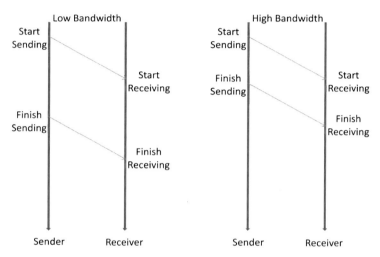

The receiver can only handle the message when all of a package has been received, which is what we call the time of finish receiving. A higher bandwidth will cause an earlier finish receiving time, because the sending time for the same length in a high bandwidth is shorter.

The distance of computer rooms and length of cable are also very important. Though light speed is very fast, 300 million meters per second, it is still not infinite, especially in the microsecond world of HFT. We can see that a 150-meter length cable will cost 1 microsecond for a round trip between the exchange and the trading system. And a longer distance may also indicate more switches are needed, which has a much larger impact on latency. High-frequency traders should always try to get the best place in computer rooms so as to get at least the same speed as the others.

If a strategy requires some market data from other exchanges, a fast dedicated line may be used for it. And in some cases, different kinds of lease lines are in the market with a latency difference in microseconds. But it is still valuable to use these better lines with higher costs for latency critical strategies. For a cross-market arbitrage, some HFT funds are looking for a middle point between the exchanges, so as to find the arbitrage opportunities earlier than the others. The competition on a network environment is just like an arms race. Every fund is looking for any methods other funds used, and afraid of being behind of others.

4.4 Core Technologies

The trading system for HFT is a highly competitive area in software design. Every designer must try his/her best to crush any potential lag from the system. We will discuss several ways to do it in the following subsections.

4.4.1 Single-Process vs. Multi-Process

A widely used design of a trading system is using the multi-process approach. The most obviously independent part is the trading interface. In some designs, risk control and the surveillance module are put in another process. All these kinds of splits may be correct in a common system design. However, for a response time critical system, it is another story. We should analyze the consequence of multi-process design.

An independent trading interface process connected by a network means you must at least add two steps of network communication in each response cycle. These two steps of communication will cause at least four operation system traps (two for sending and two for receiving), two times of marshalling, two times of un-marshalling, and two times of wake up. Therefore, such a kind of design may add at least hundreds of microseconds to the response time. This is a huge latency. Considering the current situation, about 10 microseconds of internal response time, adding a process may indicate that one has failed. There are some benefits from multi-process, such as fewer things will be impacted from changes, and it is possible to do a realtime fail over and so on. But these benefits are not so important for a trading system of HFT. Therefore, we should always use a single process if possible.

If we have to use multi-processes in a trading system, we should try our best to minimize any impact of the communication between them. The following are some considerations for it.

- Minimize the steps of communication between processes;
- Try to put all processes in the same server;
- Instead of network communication, use sharing memory if possible;
- Use spin lock to do synchronization.

4.4.2 Code Optimization

The standard software code optimization methods are still very important for the trading system. The following points are very important.

Optimization of the data structure and algorithms is always an important area. As the key benchmark for a trading system is the internal response time, any logic in this critical path should be considered carefully. Some data structures should always be used in any HFT system, such as the depth of market data, and the account order book. This data should be in a specially designed structure fitted for the real interaction requirements of this system.

We should try our best to implement all algorithms in an incremental way, avoid any large loops if possible. This is especially important in the strategy for implementation. As many strategies will use a lot of historical market data to generate signals, we should avoid recalculating everything on each new set of market data. We can always try to find some way to buffer some median result to help to quicken calculations of the final result. For example, if a trader is going to calculate the standard deviation of some market data, it is not as easy to do the incremental calculation directly for it. But if we always keep the sum and square sum of this data, which can be easily calculated incrementally, we can get the standard deviation in a constant time.

A trading system should be a message-driven system. Most of the codes are handling processes of some message. To improve the performance of messaging is quite important for response time. The following are some important tips to define such a messaging system.

- Use subscribe/listener model;
- Minimize the messages subscribed;
- Directly dispatch each message to objects listening to it;
- Keep each message handling process short;
- Use function calls to send message, no lock wake-up mechanism.

The database design is a favorite design model for many designers, even in a non-database system. It is a problem that they always like to use a primary key for reference, which will cost a tree search each time you want to get detailed information. A better design is an object-oriented design, where you must use a pointer to act as reference. In such a case, all important data structures should be fixed in memory, including instrument information, product information, market data, position information and account information. We can keep all this data in predefined structures with suitable references. Hence, any time we get some incoming message, we will use one tree search to find the corresponding object, and then you can only use a pointer to get all related objects free of any search.

4.4.3 Memory Management

A very important difference between the trading system and many other systems is that the former is not designed for running a very long time. Trading systems will not run over 24 hours, but at much shorter time in most cases. Therefore, a lot of worry of memory leak in other systems can be ignored in the trading system. The problem of trading systems with memory management is trying to improve the speed for memory allocation. To achieve this, we can use several methods as follows:

- Allocate enough space for important data structures at the very beginning;
- If you have to do dynamic allocation in the middle of running, do not do allocation one by one, but allocate a lot and manage it yourself;
- Use some simple fix-sized memory management algorithms. Avoid using complicated allocation and free function of the system;
- Do not free any memory to the system in the middle of running;
- Put all temporary usage objects in a stack, not in heap;
- Use object references as method parameters, and avoid triggering a copy constructor of objects.

4.4.4 Managing CPU Cache

Some new kinds of optimization specially based on the understanding of CPU and internal architecture have been introduced to HFT systems. These optimization methods may only be used in some specified hardware, so it is not widely discussed in theoretical research. Nevertheless, it is quite useful. One of such optimization is to improve the CPU cache hitting rate. The access time for cache and normal memory is extremely different, a system can benefit a lot by increasing cache hitting rate.

In order to increase the cache hitting rate, we must first understand how CPU cache are managed, and how process memory is managed. In most CPUs, a Least Recently Used (LRU) algorithm is used for the CPU cache. It will discard the least recently used items first. If some data is frequently used, it may be kept in a CPU cache for a long time to increase the hitting rate. The problem is that if we have a lot of frequently used data, it will still easily be discarded. Therefore, trying to fit our frequently used data to cache is the key point in this method.

To manage your CPU cache, you must plan how many bytes of data are going to be put in a cache. This should be no larger than 1/3 of the real cache size, because there must be some cache used for the program code and stack. And you need to organize your frequently used data in a continuous memory, which may lower the possibility of discarding, and may also cause it to be loaded into a cache at the same time. When running a system in such a design, you must use a dedicated CPU core to run your system, so that it will not be disturbed by switching processes.

Another consideration is the locks in a multi-thread system. Though the implementation of a lock is very simple, and even may be just several machine codes, it may cause some dirty CPU caches, and lower the cache hitting rate. Some special data structures may help solving this problem. Spin lock is another special technology for HFT. In normal systems, spin lock will waste a lot of CPU time, and may not be suggested. However, in HFT, the advantage of spin lock with no context switching is quite important for low latency programming. And the server is dedicated to the usage of HFT, if a CPU has spare time, it is just a waste. Therefore, spin lock can be used in HFT systems.

Chapter 5

Stationary Process and Ergodicity

We find that the high-frequency trading (HFT) is closely related to the ergodic theory of stationary processes [12]. As we have shown in the previous chapters, in most HFT cases, the trader cannot guarantee to make profit in each of his trades. However, if a trader can repeatedly apply the same algorithm with positive expected profit to trade, then the strong ergodic theorem will lead to a steadily increasing accumulated profit as the total sum, if the profit series forms an ergodic stationary time series. That is the main thing we will explain in the next four chapters.

Since we do not assume that the majority of our readers has a background in modern probability theory and statistics, we introduce first some basic ideas of modern probability theory and stationary processes in this chapter. We will not follow the usual mathematical trail of definition–proof–theorem for the convenience of our non-mathematical readers. For the same reason, we also avoid some precise mathematical definitions from time to time. The readers who are familiar with probability theory and statistics may skip this chapter except for the last section.

5.1 Some Basics of Probability Theory

We introduce some basic definition and facts in probability theory in this section. The readers who are not interested in mathematical details may skip this section and use it as reference for Section 5.2.

5.1.1 Probability Space

The statements

$A = \{$the final trading price of Asset J before 14:00:00 is ¥12.01$\}$,

and

$B = \{$the final trading price of Asset J before 14:00:0 is at least ¥12.00$\}$,

are examples of **random events** (ab **events**). In probability theory, we consider a collection of random events. If event A′ implies event B′, then we denote A′ ⊆ B′. In the previous example, we have A ⊆ B. We say that two events A and B are disjoint, if at most one of them happens. Denote by A^c the complement of A, i.e., the event that A does not happen. Therefore, if A is a random event, so is A^c. We denote by Ω the sure event. The complement of Ω is the null event, which never happens, denoted by ∅. We make a mathematical convention that ∅ ⊆ A ⊆ Ω for any event A.

If ω is an event such that A ⊆ ω implies that A = ∅ or A = ω, then we call ω an **elementary event**. The notation A ∩ B denotes the event that both A and B happen. We call A ∩ B the **intersection** of A and B or the **joint event** of A and B. The notation A ∪ B denotes the event that at least one of them happens. We call A∪B the **union** of A and B. If A ⊆ B^c, then we say that they are **disjoint** and denoted as A ∩ B = ∅. Let us consider an example. When one rolls a die once, there are six elementary events $\{1, 2, 3, 4, 5, 6\}$. The event A = {to get a number strictly less than 3} is not elementary but the union of elementary events {1} and {2}. The event B = {to get 5 or 6} is disjoint with A, and A ∪ B is the event {to get 1, 2, 5, 6}.

Since Ω is the sure event, Ω is the union of all elementary events. We use particularly the notation $\omega \in \Omega$ to emphasize that ω is an elementary event in Ω. Generally, if $\omega \subseteq$ A and ω is an elementary event, we denote $\omega \in$ **A** to emphasize that ω is an elementary event in A. We call Ω **the elementary space** composed of all elementary events.

In mathematics, a set A is called **countable** if all of its elements can be counted one by one in a row as $\{a_1, a_2, a_3, \ldots\}$ (which may be finite). For example, the set of all natural numbers $\{1, 2, \ldots\}$ is countable but the open interval (1,2) is not, because there are too many irrationals. With that concept, we can define further the operation of events. If $\{A_i, i = 1, 2, \ldots\}$ are events, their countable intersection

$$\cap_{i=1,2,\ldots} A_i = A_1 \cap A_2 \cap A_3 \cap \ldots$$

means the event that all of those events happens, and their countable union

$$\cup_{i=1,2,\ldots} A_i = A_1 \cup A_2 \cup A_3 \cup \ldots$$

means the event that at least one of them happens. They are all random events.

If we denote the price of an asset at time s as $P(s)$, then the collection of all possible random events (related to $P(s)$) before time t can be denoted as F_t. We use the notation $A \in F_t$ to denote that A belongs to F_t. It is easy to see that F_t satisfies the following properties:

(1) If $A_i \in F_t$, then $\cap_{i=1,2,\ldots} A_i \in F_t$;
(2) If $A_i \in F_t$, then $\cup_{i=1,2,\ldots} A_i \in F_t$;
(3) If $A \in F_t$, then $A^c \in F_t$;
(4) $\Omega \in F_t$.

In mathematics, a collection satisfying the above four properties is called a σ-**field**. We call F_t the σ-**field of random events up to time** t. It is easy to see that if $s < t$ and if $A \in F_s$, then $A \in F_t$. We denote as $F_s \subseteq F_t$. Usually, we denote F as the largest collection which contains all random events. Therefore, we have $F_s \subseteq F$ for any s. We call the increasing family $\{F_t\}$ as **the filtration of σ-fields**. The above definition is important in finance. For $s < t$, the event $[P(s) < P(t)]$ is an event in F_t but not an event in F_s, as we do not know the price $P(t)$ by time s yet.

The **probability measure P** is an assignment of a non-negative number $P[A]$ to each random event $A \in F$ such that

(a) $P[\cup_{i=1,2,\ldots} A_i] = P[A_1] + P[A_2] + \cdots = \sum_{i=1}^{\infty} P[A_i]$ if A_i are countably many mutually disjoint events;
(b) $P[\Omega] = 1$ and $P[\emptyset] = 0$.

Finally, we call such a triple (Ω, F, P) a **probability space**. In other words, a probability space is composed of elementary events, a σ-**field** F of random events, and a probability measure defined on F.

5.1.2 Random Variables

If for each elementary event ω, we assign a value $X(\omega)$ such that for each given constant c, the set $\{\omega; X(\omega) < c\}$ (i.e., the set of all ω satisfying the property "$X(\omega) < c$") is a random event in F, then we say X is a **random variable**. We can also say that X is **measurable with respect to F** and denote as $X \in F$. More generally, if G is a σ-field of random events (not necessarily coincides with F) and $[X < c] \in G$ for any given constant c, then we say that X is **G-measurable** and denote as $X \in G$. The conditional expectation $E[X|G]$ defined later will be an example of G-measurable random variable.

If $\{F_\lambda\}$ is a family of σ-fields of random events, their intersection $\cap F_\lambda$ also satisfies the Conditions (1)–(4), so $\cap F_\lambda$ is also a σ-fields of random events. Let $\{X_\kappa\}$ be a family of random variables depending on a parameter κ. The intersection of all σ-fields which contains all sets $\{[X_\kappa < c]; c, \kappa\}$ is still a σ-fields. We call it **the σ-field of random events generated by** $\{X_\kappa\}$ and denoted as $\sigma(\{X_\kappa\})$. For example, the σ-field generated by the asset price $\{P(s), s \leq t\}$ up to time t is just F_t when one only considers the random events described by the data of $\{P(s)\}$.

If there is a non-negative function $f(x)$ such that

$$P[X \in (a, b)] = \int_a^b f(x)dx,$$

we say that X is a random variable of **continuous type** and $f(x)$ is the probability density function of X. For a random variable X of continuous type, we define

$$E[X] = \int_{-\infty}^{\infty} x f(x)dx,$$

as its mean or **mathematical expectation**.

If a random variable X may assume values $x_1, x_2, \ldots$ with probability $p_1, p_2, \ldots$ respectively, we call it a **discrete random variable**. Then its mean or mathematical expectation is given by

$$E[X] = p_1 x_1 + p_2 x_2 + \cdots = \sum_{i=1}^{\infty} p_i x_i.$$

The function $(X - E[X])^2$ is still a random variable. Its mean $E[(X - E[X])^2]$ is called the **variance** of X, denoted by $\mathrm{Var}(X)$. We call $\sqrt{\mathrm{Var}(X)}$ the **standard deviation** of X.

When a continuous type random variable has density function $\frac{1}{\sqrt{2\pi}}e^{-\frac{1}{2}x^2}$, we say that X has standard Gaussian distribution (or standard normal distribution). Generally, if there are constants $\sigma > 0$ and μ such that $(X - \mu)/\sigma$ has standard Gaussian distribution, then we say X is a **Gaussian random variable** with mean μ and standard deviation σ. We can verify by integration that if X has Gaussian distribution with mean μ and standard deviation σ, then $E[X] = \mu$ and $E[|X - \mu|^2] = \sigma^2$.

5.1.3 Conditional Probability

Suppose that A and B are two random events. We denote

$$P[A|B] = P[A \cap B]/P[B],$$

and call it the **conditional probability** of A given B. For example, when one rolls a fair die, denote A as the event "to get a '5'" and B as the event "to get an odd number". Then $P[A] = 1/6$ and $P[B] = 1/2$. If we knew already the result is an odd number, then the conditional probability of getting "5" is $(1/6)/(1/2) = 1/3$.

If $B_1, B_2, \ldots$ are countably many (mutually) disjoint events, all their possible union form a σ-fields G of random events. Then, we define the conditional probability given G: when $\omega \in B_i$

$$P[A|G](\omega) = P[A|B_i],$$

which becomes a random variable.

Given an event B, the conditional probability $P[A|B]$ (when A moves) satisfies (a) and (b) in the previous definition of probability measure. So we can also define similarly the **conditional expectation** $E[X|G]$ of a random variable X given σ-field G. When $E|X|^2$ is finite, then $E[X|G]$ is the best approximation of X based on the events in G. That is, the following two properties hold:

(a) $[E[X|G] < c] \in G$ for any constant c;
(b) $E[|X - E[X|G]|^2] = \mathrm{Min}\{E|X - Y|^2; Y \in G\}$,

where $Y \in G$ means that $[Y < c] \in G$ for any constant c. The above equality can be used as a definition of conditional expectation as long as G is a σ-field of random events (does not necessarily coincide with F). **In other words, $E[X|G]$ is the best approximation of X among all G-measurable random variables.** Let us give an example. Suppose we roll a die twice and denote the results as X_1 and X_2 respectively. Then, X_1 and X_2 are random variable, so is their sum $X = X_1 + X_2$. Let G be the σ-field of events generated by X_1. Then X is not G-measurable and we can verify that

$$E[X|G] = X_1 + E[X_2] = X_1 + 3.5.$$

We call $X = (X_1, \ldots, X_n)$ an n-dimensional random vector, if each component is a random variable. Similarly, we can define the probability density function $f(x_1, \ldots, x_n)$ of X when

$$P[X_i \in (x_i, x_i + dx_i), i = 1, \ldots, n] = f(x_1, \ldots, x_n)\, dx_1 \ldots dx_n.$$

We say that X has an n-dimensional Gaussian (normal) distribution, if for any constants $a_1, \ldots, a_n$ (not all zero), $a_1 X_1 + \cdots + a_n X_{1n}$ is a Gaussian random variable.

We say that the events $A_1, A_2, \ldots$ are mutually **independent** if for any selection $A_{i(1)}, A_{i(2)}, \ldots, A_{i(n)}$ from them, the probability of the intersection is equal to the product of their probabilities, i.e.,

$$P[A_{i(1)} \cap \ldots \cap A_{i(n)}] = P[A_{i(1)}] \ldots P[A_{i(n)}].$$

It is easy to see that two events A and B are independent if and only if $P[A|B] = P[A]$. We say random variables $X_1, X_2, \ldots$ are mutually independent, if the events from different $\{\sigma(X_i)\}$ are mutually independent. If two Gaussian random variables X and Y are independent, then (X, Y) has two-dimensional Gaussian distribution.

5.1.4 Two Main Theorems

The definitions introduced in the previous subsections form the main structure of the modern probability theory. The first result one can obtain from that structure is the law of large numbers. Let us introduce it in an intuitive way. Suppose, we run a series of independent trials. We denote the possible outcomes as $X_1, X_2, \ldots$ (before the experiments) which are mutually independent discrete random variables with the same probability distribution

$$P[X_i = y_j] = p_j.$$

We denote its experimental results as $x_1, x_2, \ldots, x_n$ in n independent trials. Denote by n_j the number of trials such that the outcome $x_i = y_j$. Then, n_j/n is called the **relative frequency** that value y_j appears. Although we cannot predict the resulted series $x_1, x_2, \ldots$, our experience tells us that n_j/n will converge to the probability p_j when $n \to \infty$. Therefore, $\frac{1}{N}(X_1 + \cdots + X_N)$ will be approximately $E[X_1]$ when N is very large. Since the random variable of continuous type can be approximated by that of discrete type, the same statement should be also true. We write it as the following theorem and omit its detailed mathematical proof.

5.1.4.1 *The strong law of large number*

Suppose $X_1, X_2, \ldots$ are independent and identically distributed random variables with mean $E[X]$. Then $(X_1 + \cdots + X_n)/n$ converge to $E[X]$ with probability 1 when $n \to \infty$.

With the strong law of large numbers, we can also prove the fact that the relative frequency converges to the corresponding probability as its corollary. Indeed, we may denote by $I_j(i)$ the random variable which equal to 1 if $X_i = y_j$ and equal to 0 otherwise. Such a random variable is called as the **indicator of the event** $[X_i = y_j]$. It is easy to see that $E[I_j(i)] = p_j$ and $\{I_j(i)\}$ is an independent sequence of random variables. Moreover, $[I_j(1) + I_j(2) + \cdots + I_j(n)]/n$ is just the relative frequency. By the strong law of large numbers,

$$n_j/n = [I_j(1) + I_j(2) + \cdots + I_j(n)]/n \to E[I_j(1)] = p_j.$$

The strong law of large number is the foundation of mathematical statistics, without which the relative frequency would have no probabilistic meaning.

There is another basic theorem which is a little more difficult to prove. We list it here without a proof:

5.1.4.2 The central limit theorem

Suppose $X_1, X_2, \ldots$ are independent and identically distributed random variables with mean 0 and variance σ^2. When n is sufficiently large $(X_1 + \cdots + X_n)/\sigma\sqrt{n}$ is approximately a standard Gaussian random variable.

Mathematical statistics has been developed based on the above two basic theorems. When a statistician wants to study a target population, he draws independent samples $\{X_i\}$. Although he does not know the exact probability distribution, $(X_1 + \cdots + X_n)/\sqrt{n}$ will be approximately Gaussian when n is large and the relative frequency of an event will approach the corresponding probability of that event.

5.2 Stochastic Process

A **stochastic process** $\{X(t)\}$ is a family of random variables indexed by time t. Here, the time t can be continuous or discrete. When one observes a stochastic process continuously, one gets a function $x(t)$. We call $x(t)$ a **realization** or a **sample path** of $\{X(t)\}$. We can consider such a sample path as an elementary event ω and denote $x(t) = X(t, \omega)$. When its paths are all increasing functions in t, we say the process is an **increasing process**. When its paths are all continuous functions in t, we say the process is **continuous**.

5.2.1 Examples of Stochastic Processes

There are many examples of stochastic processes. The following two processes are the most classical ones for continuous time t:

(1) A continuous process $\{W(t)\}$ is called a standard **Brownian motion** (or Wiener process) if

    (a) $W(0) = 0$;
    (b) $W(t) - W(s)$ has Gaussian distribution $N(0, t - s)$ for any $s < t$;
    (c) for any finite time $t_1 < t_2 < \cdots < t_n$,

$$W(t_2) - W(t_1), \ W(t_3) - W(t_2), \ \ldots, \ W(t_n) - W(t_{n-1})$$

    are mutually independent.

As one of its application, the simplest model for stock price $S(t)$ is that
$$\log S(t) = \log S(0) + \sigma W(t) + ct,$$
or its equivalent version
$$S(t) = S(0)\exp\{\sigma W(t) + ct\},$$
where c and σ are constants. We call it a **geometric Brownian motion**.

(2) An increasing process $\{N(t)\}$ is called a **Poisson process** with parameter λ if

   (a) $N(0) = 0$;
   (b) $N(t) - N(s)$ has Poisson distribution with parameter λ for $t > s$:
   $$P[N(t) - N(s) = k] = [\lambda(t-s)]^k e^{\lambda(s-t)}/k!$$
   (c) for any finite time $t_1 < t_2 < \cdots < t_n$
   $$N(t_2) - N(t_1), \; N(t_3) - N(t_2), \ldots, N(t_n) - N(t_{n-1})$$
   are mutually independent.

As its application, we often assume that the number of "news" about a stock increases as a Poisson process in behavioral finance.

5.2.2 Stationary Process and Ergodic Theory

According to the strong law of large numbers, when a statistician studies an unknown distribution, he usually needs to draw many independent samples. The more samples he obtained, the more accurate he knows the distribution. Intuitively, we need also to draw many samples to study a stochastic process, as it is a time-dependent family of random variables. However, in the application, we may just have only one sample path to trace: for example, the actual historical price of one asset in the market. One cannot find another sample path for the stochastic process corresponding to that asset. Thus, we need to find more properties of the process in order to study it statistically. A basic property to impose is stationarity. A stochastic process $\{X(t)\}$ is called a **stationary process**, if for any $a > 0$ and any finite time $t_1 < t_2 < \cdots < t_n$, the random vector
$$\{X(t_1), X(t_2), \ldots, X(t_n)\},$$

has the same probability distribution as

$$\{X(t_1 + a), X(t_2 + a), \ldots, X(t_n + a)\}.$$

Intuitively speaking, when one observes a stationary process, then the distribution will not change when one shifts time. It is easy to see that, if $\{X(t)\}$ is a stationary process and $f(x)$ is a function such that $\{f(X(t))\}$ is also a stochastic process, then $\{f(X(t))\}$ is a stationary process.

In the applications, time and prices always have their minimum units, so we actually always meet discrete time processes. When we consider the discrete time processes, we can also consider another equivalent definition of stationary process. Let T be a transformation from the basic space Ω into itself. That is, for each $\omega \in \Omega$, we assign a point $T(\omega) \in \Omega$. Moreover, we assume T is **invertible**. That is, for each $\omega' \in \Omega$, we can find $\omega \in \Omega$ such that $T(\omega) = \omega'$. We denote as $T^{-1}(\omega') = \omega$. For an event $A \in F$, $T^{-1}(A)$ denotes the set $\{T^{-1}(\omega), \omega \in A\}$. We call further T a **probability-preserving mapping** if for each $A \in F$, $T(A) \in F$ and $T^{-1}(A) \in F$ and

$$P[T(A)] = P[A] = P[T^{-1}(A)].$$

Denote $T^2(\omega) = T(T(\omega))$ and similarly denote $T^n(\omega)$ as the elementary event obtained through n-fold T. Suppose $X(\omega)$ is a random vector, T is a probability-preserving mapping, then $\{X_n(\omega)\} = \{X(T^n(\omega))\}$ is a **discrete time stationary process**. From the definition, it is easy to see that a function of stationary process is still a stationary process.

After we gave the definition for a stationary processes, we come back to our basic processes introduced in Section 5.2.1. Although Poisson process and Brownian motion are not stationary, their increment processes are stationary processes. In the case of Poisson process, for any $a > 0$ and finite time $t_1 < t_2 < \cdots < t_n$, the random vector

$$\{N(t_2) - N(t_1), N(t_3) - N(t_2), \ldots, N(t_n) - N(t_{n-1})\},$$

has the same Poisson distribution as

$$\{N(a + t_2) - N(a + t_1), N(a + t_3) - N(a + t_2), \ldots, N(a + t_n) - N(a + t_{n-1})\}.$$

A repeated sequence of a same random variable $X, X, X, \ldots$ gives a trivial example of stationary process. Its sample paths are just constants. Another trivial example of stationary process can be given as following: given a symmetric random variable X, i.e., $-X$ has the same distributions as X, then the series $X, -X, X, -X, \ldots$ is also a stationary process. In both examples, the paths stay only in a part of its range. The definition of ergodicity avoided those trivial cases. A set G on the real line is called a **Borel set**, if it is contained in the smallest σ-field of subsets containing all open intervals. A stationary process $X(t)$ is called **ergodic**, if any constant $a > 0$, for any Borel set G such that $0 < P[X(t) \in G] < 1$, the following inequality holds:

$$P[X(t+a) \in G | X(t) \in G] < 1.$$

That is, almost all paths of $X(t)$ will run over the whole range. Its intuitive meaning is: a stochastic process is said to be ergodic if its statistical properties (such as its mean and variance) can be deduced from a single, sufficiently long sample (realization) of the process. The following theorem is due to George David Birkhoff (see Refs. [5, 12, 22]).

5.2.2.1 *The strong ergodic theorem*

For a discrete time stationary process $\{X(t)\}$, when $N \to \infty$,

$$\frac{1}{N}[X(1) + X(2) + \cdots + X(N)],$$

converges if $E[|X(1)|] < \infty$. Furthermore, when $\{X(t)\}$ is ergodic, the above limit is just $E[X(1)]$.

Its continuous time version is

$$N^{-1} \int_0^N X(t)dt \to E[X(0)] \text{ (when } N \to \infty\text{)}.$$

The strong law of large number (see Section 5.1.4) can be considered as a corollary to the strong ergodic theorem. Indeed, a sequence of independent identically distributed random variables $\{X(n)\}$ is a discrete time stationary process.

Let us show another example of the application of the strong ergodic theorem. Suppose that $\{X(t)\}$ is a discrete time ergodic stationary process. We want to find the conditional probability

$$P[X(t+3) > X(t+1)|X(t+1) > X(t)],$$

based only on one observed sample path of $\{X(t)\}$. For $t > 3$, define $I(t) = 1$ (if $X(t-2) > X(t-3)$) and $I(t) = 0$ (otherwise). Define $I'(t) = 1$ (if $X(t) > X(t-2)$) and $I'(t) = 0$ (otherwise). Then, it is easy to see that $\{I(t), I'(I)\}$ is a (two-dimensional) stationary process. According to the strong ergodic theorem, the relative frequency of "$X(t+1) > X(t)$"

$$[I(4) + I(5) + \cdots + I(3+N)]/N$$
$$\to P[X(t+1) > X(t)] \quad \text{(when } N \to \infty\text{)},$$

and the relative frequency of "$X(t+1) > X(t)$ and $X(t+3) > X(t+1)$"

$$[I(4)I'(4) + I(5)I'(5) + \cdots + I(3+N)I'(3+N)]/N$$
$$\to P[X(t+3) > X(t+1), X(t+1) > X(t)] \quad \text{(when } N \to \infty\text{)}.$$

Thus,

$$\frac{I(4)I'(4) + I(5)I'(5) + \cdots + I(3+N)I'(3+N)}{I(4) + I(5) + \cdots + I(3+N)}$$
$$\to P[X(t+3) > X(t+1)|X(t+1) > X(t)].$$

When we consider the value $\{V(t)\}$ of a portfolio of assets as a stochastic process, then usually it has only probability 0 to have the observed curve as its sample path $\{V(t, \omega)\}$. So it has no much mathematical meaning to do statistics along this observed curve $\{V(t, \omega)\}$ unless it is stationary. In order to overcome this difficulty, we may consider instead a stationary process $\{X(t)\}$ which is closely related to $\{V(t)\}$. Then we can do statistics along $\{X(t)\}$ based on the strong ergodic theorem. After we get some useful conclusion on $\{X(t)\}$, we may try to see what is its impact on $\{V(t)\}$. That will be our main method throughout Chapters 6–8.

The concept of ergodicity is actually from physics. In physics and thermodynamics, the ergodic hypothesis says that, over long periods of time, the time spent by a particle in some region of the phase space of microstates with the same energy is proportional to the volume of this region.

A question is raised immediately: does the central limit theorem still hold for a stationary process? A basic condition one may impose is the so-called "mixing" condition. We will not discuss this issue in this book. Thus, we will focus on the application of the strong ergodic theorem along a sample path from now on.

The previous defined stationary process is called **strongly (or strictly) stationary process** which is difficult to be verified by statistics of data. Therefore, we introduce another definition of a stationary process. The latter is a much wider definition than the former.

A **weakly stationary process** means the following two conditions are satisfied: for each $c > 0$,

$$\text{(i)} \ E[X_t] = E[X_{t+c}]; \quad \text{(ii)} \ E[X_s X_t] = E[X_{s+c} X_{t+c}].$$

It is easy to see that a strongly stationary process with finite mean and variation is a weakly stationary process.

5.2.3 Testing Stationarity

Before we go further, we introduce some basic idea of **Hypothesis Testing** in mathematical statistics. If an event A has probability less than a preselected level of significance ε (usually, set $\varepsilon = 0.05$ or 0.01), then one usually can ignore A in a single test. For example, if one roll a fair die three times, he will not think that he will get 3 as the total sum, of which the probability is only $1/216$. If he got 3 as the sum, he would think the die is not a fair one. Thus, in order to disprove a hypothesis H_0 using experiment, the mathematical statistician would show the following logic. They preselect an event A which has only a probability less than ε to happen if the hypothesis H_0 is true. If the event A happens in the trial, then H_0 looks false (and its contradiction looks correct). Because we do not expect an event with small probability would appear in a single trial, we can reject hypothesis H_0. However, if the event A does not happen, we do not reject H_0. We will use this statistical logic in the following chapters of this book. Rigorously speaking, one cannot use one sample path to "prove" that this sample path is from a weakly stationary process. However, in statistics, there are many methods to show that a sample path is unlikely from a weakly stationary process. So the logic applied by statisticians is: if one cannot show that the current sample path is unlikely from a weakly stationary process, then we just keep the weakly stationarity hypothesis. In the above example, H_0 is called the **null hypothesis** and its contradiction is called the **alternative hypothesis** denoted by H_1.

Given a sample path $\{x(t)\}$, there are several statistical software available to test $H_0 : \{x(t)\}$ is from a stationary process; $H_1 : \{x(t)\}$ is NOT from a stationary process.

A most popular one is Augmented Dickey–Fuller test. One can find it in MATrix LABoratory (MATLAB) software package. We will not discuss the details here.

5.2.4 Semi-Martingales and Filtering Problem

In many application problems, we meet a lot of random times. There are two kinds of random times classified by whether or not it can be identified at the moment that it comes. For example, such an order can be executed: to sell at "market price" immediately if Asset J reaches \$26 or above today. However, the following order cannot be executed: to sell Asset J at today's maximum price. The maximum price can be only identified by the end of today's trading, which is often too late to sell. The former is called a stopping time. Therefore, we need to introduce the definition of stopping time.

A non-negative random variable $T(\omega)$ is called a **stopping time**, if $\{T \leq t\}$ is an event in F_t for any t. As an example, if T is the time "when the price $P(t)$ of Asset A reaches or above \$26", then T is a stopping time. In fact

$$\{T \leq t\} = \cup \{P(s) \geq 26; s \leq t\} \in F_t,$$

where the union $\cup$ is taken over all the time (at which a trade is made) up to t.

For a pair of stopping times $S \leq T$, we use $[S, T)$ to write a random interval. For each elementary event ω, $[S, T)$ is the set of all t such that $S \leq t < T$. We denote by $I_{[S,T)}(t, \omega)$ its indicator. That is, $I_{[S,T)}(t, \omega) = 1$ if $S(\omega) \leq t < T(\omega)$ and $I_{[S,T)}(t, \omega) = 0$ otherwise. Then the indicator

Stationary Process and Ergodicity

of a random interval is a stochastic process. Using that indicator, one can precisely write the price process. Let $T(i)$ be the time of the i-th trade and a_i be its value, then the last trade price of an asset will be

$$S(t, \omega) = a_1 I_{[T(1), T(2))}(t, \omega) + a_2 I_{[T(2), T(3))}(t, \omega) + \cdots .$$

If $\{X(t)\}$ is a stochastic process such that $\{X(t) < c\}$ is an event in F_t for any c (denoted as $X(t) \in F_t$), then we say $\{X(t)\}$ is **adapted** to $\{F_t\}$.

Let $\{M(t)\}$ be an adapted process such that $E|M(t)|^2$ is finite for each $t > 0$. If for $s < t$,

$$E[M(t)|F_s] = M(s),$$

then we call $M(t)$ a **martingale** with respect to $\{F_t\}$. In other words, at any time, a martingale is the best approximation of its future value according to current information. The standard Brownian motion is a typical example of martingale. Indeed, $W(t) - W(s)$ has null expectation and is independent of any random event in F_s. So that $E[W(t) - W(s)|F_s] = 0$. On the other hand, since $W(s) \in F_s$, $E[W(s)|F_s] = W(s)$. Therefore,

$$E[W(t)|F_s] = E[(W(t) - W(s)) + W(s)|F_s]$$
$$= E[W(t) - W(s)|F_s] + E[W(s)|F_s]$$
$$= W(s).$$

Suppose we have a function $X(t)$ which has derivative $b(t)$ with respect to time t. Then, we can write

$$dX(t) = b(t)dt.$$

When there is a martingale $\{M(t)\}$ as "noise", we have

$$X(t) - X(0) = \int_0^t b(s)ds + M(t), \tag{5.2.1}$$

or denoted formally as

$$dX(t) = b(t)dt + dM(t).$$

Given a filtration $\{F_t\}$, $Y(t, \omega)$ is a **semi-martingale** if there exists three $\{F_t\}$-adapted processes such that

$$Y(t, \omega) = A_+(t, \omega) - A_-(t, \omega) + M(t, \omega),$$

where M is a martingale, both $A_+(t, \omega)$ and $A_-(t, \omega)$ are increasing in t for fixed ω. We denote $A(t, \omega) = A_+(t, \omega) - A_-(t, \omega)$. A process which can be written as a difference of two increasing process is called **a process of bounded variation**. Thus, $\{A(t, \omega)\}$ is a process of bounded variation. We have

$$Y(t, \omega) = A(t, \omega) + M(t, \omega). \qquad (5.2.2)$$

We introduce the notation $b^+(t) = \text{Max}\{b(t), 0\}$ (i.e., $b^+(t) = b(t)$ when $b(t)$ is positive and $b^+(t) = 0$ otherwise) and $b^-(t) = \text{Max}\{-b(t), 0\}$ (i.e., $b^-(t) = |b(t)|$ when $b(t)$ is negative and $b^-(t) = 0$ otherwise). Then (5.2.1) becomes

$$X(t) - X(0) = \int_0^t b(s)ds + M(t) = \int_0^t b^+(s)ds - \int_0^t b^-(s)ds + M(t).$$

Thus $X(t)$ is also a semi-martingale.

The general idea of **filtering problem** in signal processing is to form some kind of "best estimate" for the true value $\{b(t)\}$ of the above system, given only some (potentially noisy) observations of that system. The problem of optimal non-linear filtering was solved by R.L. Stratonovich. The solution, however, is infinite-dimensional in the general case. Certain approximations and special cases are well-understood: for example, the linear filters are optimal for Gaussian random variables, and are known as the Wiener filter and the Kalman–Bucy filter.

5.2.5 Stationary Process as Noises

We consider a different problem in HFT. We assume that the price $P(t)$ of certain portfolio has the decomposition

$$P(t) = A(t) + Z(t), \qquad (5.2.3)$$

where the "noise" $Z(t)$ is a stationary process, which is not a martingale. $A(t)$ moves much slower than $Z(t)$. Contradictory to the filtering problem, we do not care about $A(t)$ but $Z(t)$ in HFT. From probability theory, we know that a martingale $M(t)$ being stationary is equivalent to $M(t) \equiv M(0)$. So the martingale in (5.2.2) is not stationary unless it has constant sample paths. In order to decompose $\{Y(t)\}$ in (5.2.2) into two parts as the right-hand side of (5.2.3), we need to find a process $Q(t)$ such that

$$Y(t) = [A(t) - Q(t)] + [Q(t) + M(t)],$$

and $[Q(t) + M(t)]$ is a stationary process which moves much faster than $[A(t) - Q(t)]$.

Let us explain what we need heuristically now. A precise example of application will be given in the last section. Statistically, we need to find an interval $[a, b]$ such that $Z(t)$ cross this interval fast and frequently while the increments of $A(t)$ is bounded by a small fraction k of $b-a$. One buys when $Z(t)$ is less than a and sells when $Z(t)$ is above b. If the transaction cost is lower than $(1-k)(b-a)$, then the profit will be positive. A good HFT algorithm should have its profit at each trade to form a stationary process with positive mean. Then the strong ergodic theorem gives us cumulative total profit. In HFT, the profit of each trade is very small, at the level of 0.01%. However, if one trade 30–40 times per day, the daily profit will be 0.3–0.4% and the annualized gain will be around 100%.

5.3 Time Series Analysis

Our discussion in the previous section is close to the time series analysis in statistics. A **time series** is a sequence of data points, measured typically at successive points in time spaced at uniform time intervals. The sample path of a discrete timed stochastic process is certainly a time series. When one records a sample path of a continuous-time process, and when one does actual computation related to a continuous-time process, all the numerical data has to have minimum units which are discrete. Therefore, in the applications, one uses more time series than continuous-time process. Although the later gives better analytic tools, we explain some basic methods of time series in this section.

Since the stationary property is important when one only has a realized sample path to analyze, we need to transfer non-stationary data into stationary one. The following method is frequently used by statisticians. First step is to plot the data. Inspection of the graph may suggest the possible decomposition of data $X(n)$ as

$$X(n) = m(n) + S(n) + \varepsilon(n), \tag{5.3.1}$$

where $m(n)$ is a slowly changing deterministic function known as a "**trend component**", $S(n)$ is a deterministic function with period d referred to as a "seasonal component" such that $S(n) = S(n+d)$ and that $S(1) + \cdots + S(d) = 0$), and $\varepsilon(n)$ is a "random noise component" which is weakly stationary and $E[\varepsilon(n)] = 0$. We need to emphasize here, in practice, the only observable data are $\{X(n)\}$. None of the three components in (5.3.1) is observable. Our next step is to eliminate the seasonal component.

We introduce the **lag-d difference operator** ∇_d defined by

$$\nabla_d X(n) = X(n) - X(n-d) \quad \text{(for } n > d - 1\text{)}.$$

When $d = 1$, we just denote $\nabla_1 = \nabla$ and call it the difference operator. Since S has period d, $\nabla_d S(n) = 0$. Thus, (5.3.1) becomes

$$\nabla_d X(n) = \nabla_d m(n) + \nabla_d \varepsilon(n). \tag{5.3.2}$$

Equation (5.3.2) can be rewritten as a new time series with no seasonal component:

$$X(k) = m(k) + \varepsilon(k). \tag{5.3.3}$$

There are a few methods to eliminate $m(k)$ in time series analysis. When $m(k)$ is approximately linear over interval $[k-q, k+q]$ and the average of error terms over this interval is close to zero (when q is large, thanks to the law of large number), a well-known method is to use two-sided moving average:

$$m'(k) = (2q+1)^{-1}[(X(k-q) + X(k-q+1) \\ + \cdots + X(k+q-1) + X(k+q)].$$

Then $\varepsilon'(k) = X(k) - m'(k)$ may have a chance to be stationary. After that, we need to test the stationary. There are quite a few methods and software available. If one cannot reject the stationarity of $\varepsilon'(k)$, then we may assume it is stationary. We will not discuss those various computer programs here.

We should mention that the above method of eliminating $m(k)$ is not useful in our application to HFT. In HFT, we can only use current information (events in F_k) to predict the near future. The data $X(k+1), \ldots, X(k+q)$ are not available at time k.

Equation (5.3.3) has some similarity to (5.2.3) but not the same, as we did not assume that $\{A(t)\}$ is deterministic in (5.2.3).

For a weak stationary time series $\{X(n)\}$, the mean $E[X(n)]$ is a constant. If we denote $E[X(n)] = C$, then the covariance of $X(n)$ and $X(n+h)$ is given by

$$\Gamma(h) = E[(X(n) - C)(X(n+h) - C)].$$

We call

$$\rho(h) = \Gamma(h)/\Gamma(0),$$

the **autocorrelation function** (ACF). Many authors have found that the high-frequency trading data of stocks have a significant negative $\rho(1)$ if they treat high-frequency data in the short term as a stationary time series. That means the trading price has a pulling back tendency. We will show in the next chapter that the logarithmic price $p(t)$ of CSI 300 futures has stationary increments $X(t) = p(t) - p(t-1)$. When we used 0.5 as the time unit, the ACF of $\{X(t)\}$ is as follows.

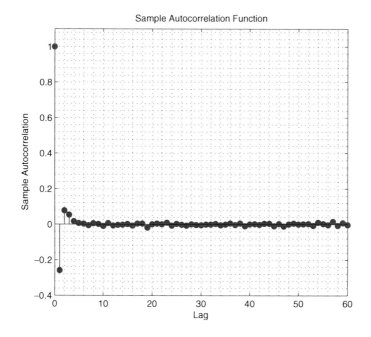

5.4 Pair-Trading Revisited

We have discussed in Section 3.2 a few examples of arbitrage. For statistical arbitrage, there are a few puzzles still remaining. Here are three of them:

(a) How frequently the arbitrage opportunity will appear per day?
(b) What is the risk that the orders are only partially executed?
(c) What is the (theoretically) expected profit?

We reconsider the arbitrage discussed in Section 3.2 from probability point of view. The following example was initially suggested by Shujin Wu [30]. Wu's original method was for pair-trading of futures of copper.

Cointegration is a statistical property of time series. Two or more time series are cointegrated if they share a common trend component proportional their values (see Eq. (5.33)) and the linear combinations of their random noise components are still random noises. Therefore, if two assets price $X(t)$ and $Y(t)$ are cointegrated, one can find a constant c such that

$$[X(t) - cY(t)] - [X(0) - cY(0)] = \xi(t),$$

where the right-hand side time series $\xi(t)$ is stationary and with mean 0. Therefore, theoretically, one can trade this combination when $\xi(t)$ is low and sell when $\xi(t)$ is high to take profit. However, some problem still exists: how long one should wait from the low of $\{\xi(t)\}$ to its high? Based on an earlier result of Shujin Wu [30], Shi Chen suggested a method (see Ref. [7] for details) to accelerate the profit process as the following.

ETF50 and ETF180 are two ETFs which track Shanghai Stock Exchange Index 50 (SSE 50) and Index 180 (SSE 180) respectively. Denote by $X(t)$ the ETF180 and by $Y(t)$ the ETF50. Since the price of ETF180 is 3–4 times of ETF50, we consider the following process

$$a(t)X(t) - b(t)Y(t) = m(t) + s(t)M(t),$$

where $a(t)$ and $b(t)$ are the positive integer-valued matching coefficients (the unit of t is 15 seconds) given by the following table.

Branch Multiplier of 180ETF & 50ETF Arbitrage Portfolio

t < 2,401	2,493	6,152	6,823	11,453	12,005	41,265	42,127	44,282	44,966
$a(t) = 4$,	3,	4,	3,	4,	3,	4,	3,	2,	3
$b(t) = 1$,	1,	1,	1,	1,	1,	1,	1,	1,	1

$m(t)$ is the longer-term trend, $s(t)$ is the standard deviation of residual, and $M(t)$ is the standardized residual. When $a(t)$, $b(t)$, $m(t)$ and $s(t)$ all are constants, the above model is reduced to a kind of two-variable co-integration model.

We use 1 second as the unit of horizontal time axis and obtain the following:

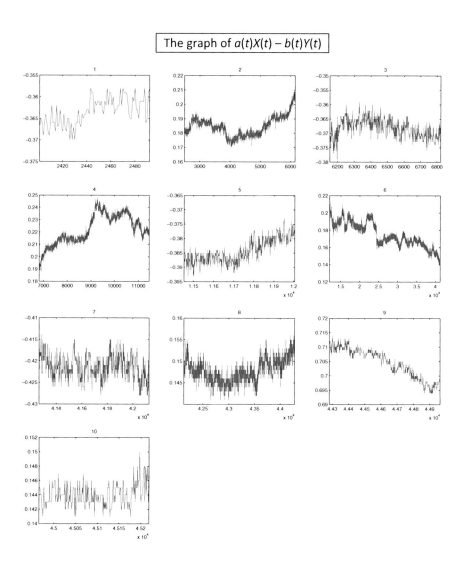

The graph of M(t)

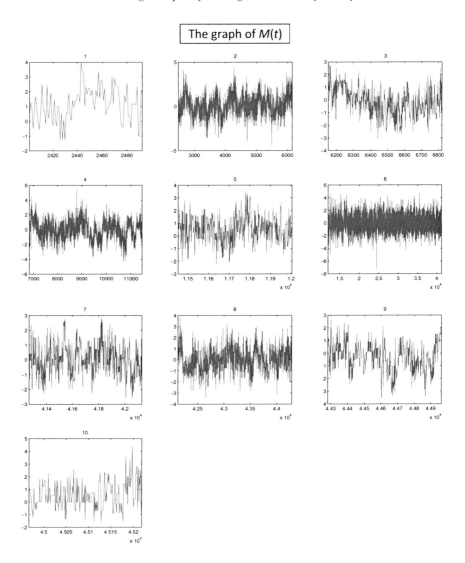

$\{M(t)\}$ is a stationary process which is more regular and more frequently reaches its high and low to compare with the original $a(t)X(t) - b(t)Y(t)$. If we buy and sell this portfolio according to the low and high of $\{M(t)\}$, we can get statistical arbitrage.

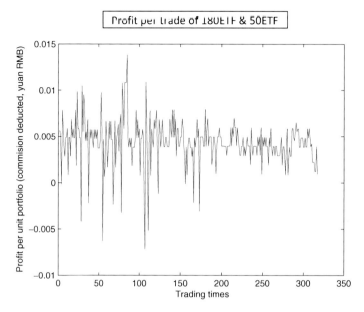

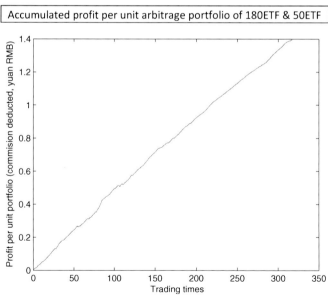

Based on this model, we find a theoretical algorithm of statistical arbitrage for HFT. As its application, this method can be used to trade the three major ETFs (ETF50, ETF180 and ETF300) in Chinese financial markets.

In the above example of HFT, one may select to trade repeatedly a basket of assets. For example, at time t_1, bid price x_1 for asset A and ask price y_1 for asset B. Then there may be four outcomes:

(a) Both order executed;
(b) Bought A at price x but failed to sell B;
(c) Failed to buy A but sold B at price y;
(d) Neither order was able executed.

Case (d) has no second step. All the others have the second step and consequences. For Case (a), we may choose time $t_2 > t_1$ to ask price x_2 for asset A and bid price y_2 for asset B. In order to make profit, $x_2 - x_1 + y_1 - y_2$ has to be strictly larger than the cost of trading. Then we may also have four cases:

(aa) Both order executed;
(ab) Sold A at price x_2 but failed to buy B at y_2;
(ac) Failed to sell A at x_2 but bought B at price y_2;
(ad) Neither order was able to executed.

Only in the Case (aa), one have a successful trade. In Case (d), one has neither loss nor profit. All the other cases lead to unpredictable results.

In the above example, we used the difference of last trade prices as the price of the pair. As we pointed out in Section 3.2.3, the chance of success is not big when we use such a pair of prices to give order. In order to increase the chance of success, one needs not only the fastest speed of an IT system but also to combine the above method with the technical analysis of both prices, which will be discussed in Chapter 6.

As we have mentioned in Chapter 1, the more assets in one's HFT basket, the more risky that his order will be only partially executed. Therefore, a problem is raised naturally: can we get statistical arbitrage just on one asset by repeatedly using technical analysis? That will be our main problem in Chapters 6–8.

Chapter 6

Stationarity and Technical Analysis

There are basically two methods to analyze the asset price: (i) fundamental analysis and (ii) technical analysis. **Fundamental analysts** examine earnings, dividends, new products, potential of the industry and the like. **Technical analysts** use charts to identify price patterns and market trends in financial markets and attempt to exploit those patterns. Technicians study technical indicators, momentum, and look for forms such as lines of support, resistance, channels and more obscure formations such as flags, pennants, balance days and cup and handle patterns. Contrasted with fundamental analysis, technical analysis holds that the prices already reflect all such trends before investors are aware of them. Uncovering those trends is what technical indicators are designed to do, imperfect as they may be. Some traders use technical or fundamental analysis exclusively, while others use both types to make trading decisions. Technical analysts do not attempt to measure a security's intrinsic value, but instead use various technical indicators to identify patterns that can suggest future activity.

We think that both fundamental and technical analyses are important. However, the former is more important for the guidance of macroscopic trading activities. When the time unit is the seconds, the former will not change much. Thus, in high-frequency trading (HFT), technical analysis is very important.

In this chapter, we use mathematical statistics to examine a few well-known technical indicators of the main contract of CSI 300 futures. We show that the logarithmic returns of main contract of the CSI 300 futures form a stationary process in each trading day and most of the popular technical indicators are, or can be transformed into functions of the logarithmic returns. Since the functions of strong stationary process are still stationary, our processes obtained from technical indicators are all stationary. Those stationary processes can be considered as revisions of the original technical indicators when the latter is not stationary. We may consider the logarithmic return together with those stationary processes as a high-dimensional stationary process. Then one can do statistics on the effectiveness of certain patterns believed in the technical analysis with the support of the strong ergodic theorem (Section 5.2.2).

In the following sections, all our statistical tests are only for **NOT REJECTING** the null hypotheses that the corresponding time series are **weakly** stationary. Nevertheless, our real hypothesis is that the time series are **strongly** stationary. Thus, there is a gap between the theory and statistical tests. However, this gap is not very critical in application. Stationarity is used mainly for the strong ergodic theorem. From our real data, the curves we are going to test are just like the textbook examples of strongly stationary process.

There have been quite a lot of statistics on the efficiency of technical indicators. However, in order to do statistics for a time series, the stationarity of data is a premium, based on which one may apply the strong ergodic theorem to deduce that the observed relative frequency converge to the corresponding probability. That is just like in the sample survey, the independence of samples is a necessary condition in order to use relative frequency to replace the probability by quoting the law of large number. As we mentioned in Chapter 5, the historical trading data of an asset forms only a sample path of a stochastic process. A particular sample path of a stochastic process has only null probability to appear. Therefore, the stationarity of data is a necessary condition if one wants to quote relative frequency of some event along one sample path as the evidence of the corresponding probability — at least from the mathematical point of view. Thus, our result gives a mathematical background to the technical analysis.

As a co-product, our results in this chapter show that a trader should not only consider the technical indicator, but also the stationary processes associated to them. The latter may be more important from the statistical point of view.

6.1 Technical Analysis

The principles of technical analysis are derived from hundreds of years of financial market data. Some aspects of technical analysis began to appear in Joseph de la Vega's account of the Dutch market in the 17th century. In Asia, technical analysis is said to be a method developed by Homma Munehisa during early 18th century which evolved into the use of candlestick techniques, and is today a technical analysis charting tool. In the 1920s and 1930s, Richard W. Schabacker published several books [23, 24] which continued the work of Charles Dow and William Peter Hamilton in their books *Stock Market Theory and Practice* and *Technical Market Analysis*. In 1948, Robert D. Edwards and John Magee published *Technical Analysis of Stock Trends* [11] which is widely considered to be one of the seminal works of the discipline. It is exclusively concerned with trend analysis and chart patterns and remains in use to the present. As is obvious, early technical analysis was almost exclusively the analysis of charts, because the processing power of computers was not available for statistical analysis. Charles Dow reportedly originated a form of point and figure chart analysis.

Dow theory is based on the collected writings of Dow Jones co-founder and Editor Charles Dow, and inspired the use and development of modern technical analysis at the end of the 19th century. Other pioneers of analysis techniques include Ralph Nelson Elliott, William Delbert Gann and Richard Wyckoff who developed their respective techniques in the early 20th century. More technical tools and theories have been developed and enhanced in recent decades, with an increasing emphasis on computer-assisted techniques using specially designed computer software.

Technical analysis seems to us is some addition to modern portfolio theory. The efficacy of both technical and fundamental analysis is disputed by the **Efficient-Market Hypothesis** (**EMH**) which asserts that markets are "efficient" (or no "arbitrage"). In consequence of this, one cannot consistently achieve returns in excess of average market returns on a risk-adjusted basis, given the information available at the time the investment is made. Many people believe that market efficiency is a simplification of the world which may not always hold true, and that the market is practically efficient for investment purposes for most individuals.

There are many techniques in technical analysis. Adherents of different techniques (for example, candlestick charting, Dow Theory, and Elliott wave theory) may ignore the other approaches, yet, many traders combine elements from more than one technique. Some technical analysts use subjective judgment to decide which pattern(s) a particular instrument reflects at a given time and what the interpretation of that pattern should be. Others employ a strictly mechanical or systematic approach to pattern identification and interpretation. The following graph shows a typical technical analysis chart. The chart shows the every 50 second data of IF1303 on February 28, 2013, which was the main contract of CSI 300 futures at late February and early March of 2013. On the chart, 10 and 20 minutes moving average (MA) are shown together with Bollinger Bands. The three rows underneath the main chart show Moving Average Convergence–Divergence (MACD), Rate of Change (ROC) and Relative Strength Index (RSI), which are all popular technical indicators.

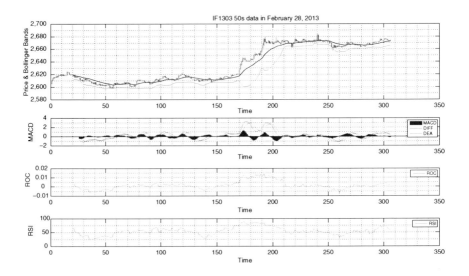

There have been a lot of mathematical and statistical models for asset prices. However, the more complicated the model is, the more difficult to verify in practice. We do not attempt to establish a new model for the price process in this book. We just use the common character implied in most models: the logarithmic returns of price are stationary. We found that for the frequently traded financial derivatives, e.g., the futures for stock index, the stationarity of logarithmic returns is justified. In the past eight years, we used the theory of stationary processes to discuss technical indicators in the Seminars of Probability, East China Normal University at Shanghai [14, 16, 19, 20, 31–33]. We found that most of those popular technical indicators either are or can be transformed into functions of the logarithmic returns of asset prices. Thus, the logarithmic returns of asset price together with the technical indicators form a multidimensional stationary process. Thus, a trader can determine his hedge based on the observed relative frequency of designed pattern according to historical data. The strong ergodic theorem tells us the observed relative frequency converge to the corresponding probability. That is the main purpose of this chapter.

From our point of view, the market background of using technical analysis in HFT is that the fundamental factors of a popularly traded security asset will not change often in a day. For example, if a traditional investment company decided to increase its holding of one asset, it will not change its decision in a few hours (which might be too late after all). Nevertheless, the microscopic technical indicators may change many times within just an hour. If one can find a stationary strategy with positive mean, then the strong ergodic theorem will give a statistically stable accumulated profit, which we will prove in Chapter 7. However, we should emphasize that we do not attempt to give evidences to support technical analysis at macroscopic level, which is affected by various news (or rumors) and relatively easier to be manipulated by institutional investors. Although, some professional investors also used HFT to illegally manipulate the market and made profit, which are not the subject to discuss in this book. Thus, we only consider the Index futures trading here, which is not easy to be manipulated.

We also would like to point out that HFT is like a card game with many players. Any strategy should be adjusted according to the change of other players. However, there are around 700,000 half-second data per month, which is enough for verifying certain stationarity in order to apply the strong ergodic theorem.

6.2 Logarithmic Return is Stationary

The main advantage of logarithmic returns is that the continuously compounded return is symmetric, while the arithmetic return is not: positive and negative percent arithmetic returns are not equal. This means that an investment of $100 that yields an arithmetic return of 50% followed by an arithmetic return of -50% will result in $75, while an investment of $100 that yields a logarithmic return of 50% followed by a logarithmic return of -50% it will remain $100.

In this chapter, we will use the price $P(t)$ of the main contract of CSI 300 futures (March 1, 2012–February 28, 2013) unless specified. We would like to emphasize that our tests are based on each trading day throughout of the whole book, as the data of different days may have jumps. Moreover, since the trading of the first 15 minutes and the last 15 minutes of each trading day is quite volatile, we will only use the data from the middle four hours of each trading day. We will always treat the data in this way throughout the remaining part of the book. Let

$$p(t) = \log P(t).$$

For any positive a, $\Delta_a(t) = p(t) - p(t-a)$ is the logarithmic return for the period a. We used statistical software to test the case where $a = 0.5$ and 5 seconds for every trading day. They all past the tests:

$$H_0 : \{\Delta_a(t)\} \text{ is stationary}; \quad H_1 : \{\Delta_a(t)\} \text{ is not stationary}$$

The following is the curve of logarithmic return when the time unit is 0.5 second and $a = 1$:

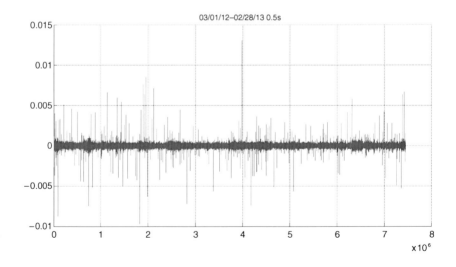

The following is the curve of logarithmic return when the time unit is 5 second and $a = 1$:

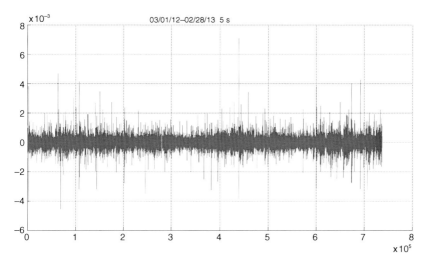

Quite a few financial researchers do their statistics on the price curve $\{P(t)\}$ which is not stationary. That is just like someone draws samples for statistics without taking care of the independency of his samples. From mathematical point of view, the relative frequency makes sense only on stationary data to which one can apply the strong ergodic theorem and use relative frequency to approximate the corresponding probability.

6.3 Moving Average and Exponential Moving Average

Denote an **n-period MA** of a discrete time process $Y(t)$ by

$$MA[Y, n](t) = \frac{Y(t) + Y(t-1) + \cdots + Y(t-n+1)}{n}.$$

When $Y(t)$ is the price $P(t)$ of the financial derivative, we simply denote $MA(t) = MA[Y, n](t)$.

Neither $P(t)$ nor $MA(t)$ is stationary. Therefore, we introduce a stationary process in order to discuss their relation statistically. Denote

$$X(1, t) = MA(t)/P(t)$$
$$= \frac{\left(1 + \frac{P(t-1)}{P(t)} + \frac{P(t-2)}{P(t)} + \cdots + \frac{P(t-n+1)}{P(t)}\right)}{n}$$

$$= \frac{[1 + \exp\{p(t-1) - p(t)\} + \exp\{p(t-2) - p(t)\} + \cdots + \exp\{p(t-n+1) - p(t)\}]}{n}.$$

Then $P(t)$ up-cross MA is equivalent to $X(1, t)$ down-crossing 1. $X(1, t)$ is a function of the logarithmic return $\{\Delta p(t)\}$. When $\{\Delta p(t)\}$ is strongly stationary, so is $\{X(1, t)\}$. Since we cannot actually prove that $\{\Delta p(t)\}$ is (strongly) stationary, we need to test the stationarity of $X(1, t)$. Our test result is that we cannot reject the null hypothesis that $X(1, t)$ is stationary. In the following discussions, we will consider more technical indicators and associate with more functions of $\{\Delta p(t)\}$. We will use all available statistical software to do stationarity tests. However, for saving our readers' time, we will not repeatedly claim that they passed the test to be stationary.

The graph of CSI 300 futures IF1303 together with its 10 and 20 minutes MA on February 28, 2013 has been shown already in Section 6.1. When $n = 20$ and the data used is per 0.5 second, the following is the graph of $X(1, t)$:

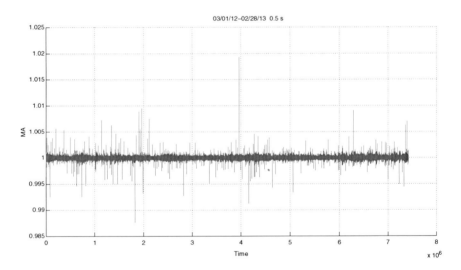

and the following is the graph when the data used is per 5 seconds:

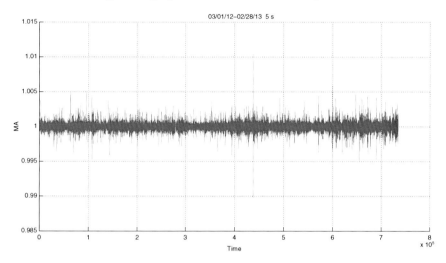

MA has a shortcoming of not reacting fast enough the price change. So we define further the n-period **Exponential Moving Average** (EMA) of time series $Y(t)$ by

$$EMA[Y, n](t) = \frac{nY(t) + (n-1)Y(t-1) + \cdots + Y(t-n+1)}{\frac{n(n+1)}{2}}.$$

The advantage of EMA is that it gives more weight on the near data. If $Y(t)$ is constant, then $EMA[Y, n](t)$ is just that constant. We simply denote $EMA(t) = EMA[P, n](t)$ when there is no possible confusion. Denote

$$X(2, t) = \frac{EMA(t)}{P(t)}.$$

Similarly to MA, $X(2, t)$ also can be written as a function of $\{\Delta p(t)\}$. Therefore, $X(2, t)$ will be stationary if $\{\Delta p(t)\}$ is strongly stationary. When $n = 20$ and the data used is per 0.5 second, the following is the graph of $X(2, t)$:

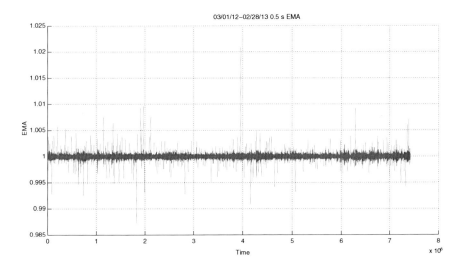

and the following is the graph when the data used is per 5 seconds:

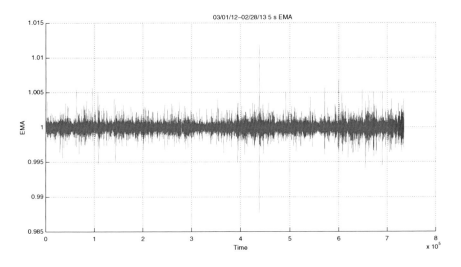

The following graph shows IF1303 together with its 10 and 20 minutes EMA on February 28, 2013.

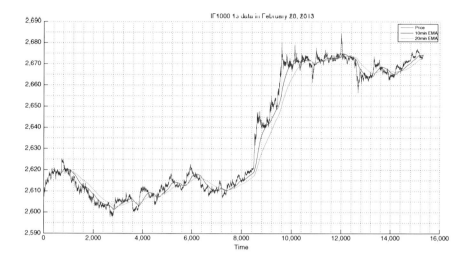

There is another way to define the **EMA**:

$$EMA[Y, m](t) = \frac{2}{m+1} \sum_{k=0}^{t} \left(\frac{m-1}{m+1}\right)^k Y(t-k).$$

When $Y(t) = P(t)$, we simply denote $\underline{EMA}[P, m] = \underline{EMA}(t)$. There are a few shortcomings in this definition:

(1) When $Y(t) \equiv c$, $\underline{EMA}[Y, m](t) = \left(\frac{2c}{m+1}\right) \sum_{k=0}^{t} \left(\frac{m-1}{m+1}\right)^k = c\left(1 - \left(\frac{m-1}{m+1}\right)^{t+1}\right)$, which is not really a time average. However, the error is ignorable when t is large;

(2) In its definition, the data before m-period are entered. However, we may truncate the sum to n terms. When n is sufficiently large, the error is ignorable.

However, $\underline{EMA}(t)$ has the following iteration formula

$$\underline{EMA}(t) = \frac{2}{m+1} P(t) + \frac{m-1}{m+1} \underline{EMA}(t-1),$$

which is convenient for fast computation. As we have mentioned before, in HFT, people appreciate the speed even at the level of microseconds. We will come back to this definition in Section 7.2 again.

If one truncate the terms in the definition of $\underline{EMA}(t)$ up to a fixed number, then $P(t)/\underline{EMA}(t)$ can also be written as a function of $\{\Delta p(t)\}$.

Therefore, the (asymptotic) stationarity will follow the strong stationarity of $\{\Delta p(t)\}$.

In technical analysis, MA levels can be interpreted as support in a rising market, or a resistance in a falling market. That interpretation can be easily re-described in terms of $\{X(1, t)\}$ or $\{X(2, t)\}$. Since $MA(t)/P(t)$ and $EMA(t)/P(t)$ are stationary, we can do statistics on those stationary ratios to study the efficiency of $\{MA(t)\}$ and $\{EMA(t)\}$.

When a MA is up-crossed by a shorter term MA, one calls it **golden cross**. When a MA is down-crossed by a shorter term MA, one calls it **death cross**. Suppose $m < n$, then the ratio

$$\frac{MA[P, m](t)}{MA[P, n](t)}$$

$$= \frac{\left(\frac{P(t)+P(t-1)+\cdots+P(t-m+1)}{m}\right)}{\left(\frac{P(t)+P(t-1)+\cdots+P(t-n+1)}{n}\right)} = \frac{\left(\frac{P(t)+P(t-1)+\cdots+P(t-m+1)}{mP(t-n+1)}\right)}{\left(\frac{P(t)+P(t-1)+\cdots+P(t-n+1)}{nP(t-n+1)}\right)}$$

$$= \frac{n}{m} \frac{\exp\{p(t) - p(t-n+1)\} + \exp\{p(t-1) - p(t-n+1)\} + \cdots + \exp\{p(t-m+1) - p(t-n+1)\}}{\exp\{p(t) - p(t-n+1)\} + \exp\{p(t-1) - p(t-n+1)\} + \cdots + 1},$$

is also a function of $\{\Delta p(t)\}$. From the strong stationarity of $\{\Delta p(t)\}$, we deduce that the above ratio is a stationary process as well. Therefore, we can do the statistics on this stationary ratio to test if the "golden cross" and "death cross" are correct for HFT.

6.4 Bollinger Bands

The **Bollinger Bands** is a technical analysis tool invented by John Bollinger in the 1980s. In 2006, we proved that (see Refs. [18] and [19]) for the geometric Brownian motion model of stock price, one can associate to Bollinger bands a stationary process ($\{X(3, t)\}$ defined below). Therefore, one can do statistics on that stationary process under geometric Brownian motion model. We discovered that the asset price indeed stays within the Bollinger bands most of the time. However, we also find that one will not get profits if one simply repeatedly bought at lower band and sold at upper band. Xuedong Huang [14] and Song Xu [31] extended the above result. We show here that the process we associated to Bollinger bands in [19] is still stationary as long as the logarithm price is strongly stationary.

Bollinger bands **consist of**:

- an n-period EMA;
- an upper band at K times an n-period standard deviation σ above the EMA ($EMA + K\sigma$);
- a lower band at K times an n-period standard deviation σ below the EMA ($EMA - K\sigma$).

Typical values for n and K are 20 and 2.5 respectively. Usually, the same period is used for both the middle band and the calculation of standard deviation.

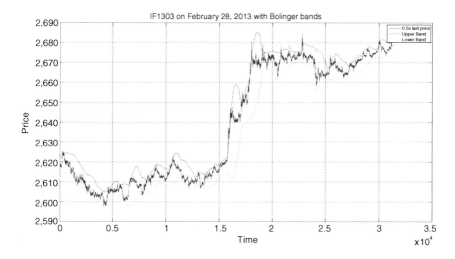

We can associate Bollinger bands with the following

$$X(3, t)$$
$$= \frac{P(t) - EMA(t)}{\sigma}$$
$$= \frac{e^{p(t)} - 2(ne^{p(t)} + (n-1)e^{p(t-1)} + \cdots + e^{p(t-n+1)})/n(n+1)}{\sqrt{\frac{1}{n-1}(e^{p(t)} - MA(t))^2 + \cdots + \frac{1}{n}(e^{p(t-n+1)} - MA(t))^2}}$$

$$= \frac{\begin{array}{c}e^{p(t)-p(t-n)}-2(ne^{p(t)-p(t-n)}+(n-1)e^{p(t-1)-p(t-n)}\\+\cdots+e^{p(t-n+1)-p(t-n)})/n(n+1)\end{array}}{\sqrt{\begin{array}{c}\frac{1}{n-1}(e^{p(t)-p(t-n)}-\frac{1}{n}(e^{p(t)-p(t-n)}+\cdots+e^{p(t-n+1)-p(t-n)}))^2+\cdots\\+\frac{1}{n-1}(e^{p(t-n+1)-p(t-n)}-\frac{1}{n}(e^{p(t)-p(t-n)}+\cdots+e^{p(t-n+1)-p(t-n)}))^2\end{array}}}.$$

The above ratio is a function of $\{e^{p(t-i)-p(t-n)}; i = 1, \ldots, n - 1\}$. When $p(t)$ is of strongly stationary increment, that ratio is stationary as well.

The stock price being between the Bollinger bands is equivalent to the inequality

$$-K < X(3, t) < K.$$

We take $n = 2,000$ in the following graph based on every 0.5 second data of $X(3, t)$ for every 0.5 second associated to $P(t)$ which is the CSI 300 futures between March 1, 2012 and February 28, 2013. $X(3, t)$ is stationary for the middle 4 hours of each trading day in that period.

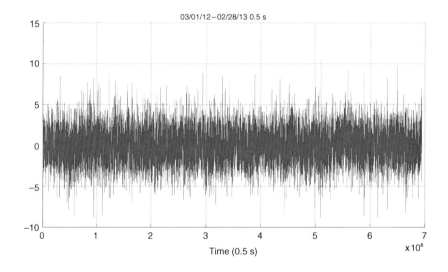

The following is the graph for every 5 seconds and $n = 200$:

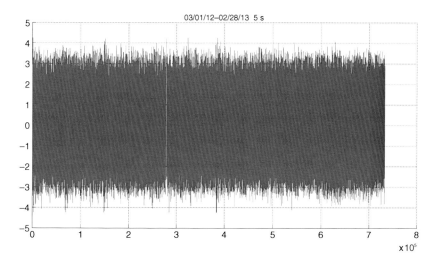

Remark: Bollinger bands now are a term trademarked by J. Bollinger in 2011.

6.5 Moving Average Convergence–Divergence

MACD indicator is one of the simplest and most effective momentum indicators. MACD was invented by Gerald Appel in the 1970s. Thomas Aspray added a histogram to the MACD in 1986 as an indicator of important moves for the underlying security. The MACD subtracts the longer MA from the shorter MA. As a result, the MACD offers the best of both worlds: trend following and momentum. The MACD fluctuates above and below the zero line as the MAs converge, cross and diverge. Traders can look for signal line crossovers, centerline crossovers and divergences to generate signals. However, since the MACD is unbounded, it is not particularly useful to identify overbought and oversold levels.

Mathematically: for $m < n$,

- $\text{DIFF}(t) = \textit{EMA}[\text{stockPrices}, m] - \textit{EMA}[\text{stockPrices}, n]$
$= 2(mP(t) + (m-1)P(t-1) + \cdots + P(t-m+1))/m(m+1)$
$-2(nP(t) + (n-1)P(t-1) + \cdots + P(t-n+1))/n(n+1)$

- $DEA(t) = EMA[MACD, k]$
- $MACD(t) = \text{Diff}(t) - DEA(t)$.

MACD turning to positive and increase signals that asset price has a tendency to increase in general situation. The period for the EMAs on which an MACD is based can vary, but the most commonly used parameters involve a faster EMA of $m = 12$ units of time, a slower EMA of $n = 26$ units of time, and the signal line as a $p = 9$ units of time EMA of the difference between the two. It is written in the form, *MACD* (faster, slower, signal) or in this case, $MACD(m, n, k) = MACD(12, 26, 9)$.

We use mathematical formula to write a function of $\{\Delta p(t)\}$:

$$X(4, t) = \frac{MACD(t)}{P(t - n - k)}$$

$$= \frac{MACD(t)}{e^{-p(t-n-k)}}$$

$$= 2(kZ(t) + (k-1)Z(t-1) + \cdots + Z(t-k+1))/k(k+1),$$

where

$$Z(t) = \frac{2}{m(m+1)} \sum_{i=0}^{m-1}(m-i)e^{p(t-i)-p(t-n-k)}$$

$$- \frac{2}{n(n+1)} \sum_{i=0}^{n-1}(n-i)e^{p(t-i)-p(t-n-k)},$$

with $m = 12, n = 26$ and $k = 9$. $Z(t)$ is a function of $\{p(t-i) - p(t-n-k); i = 1, \ldots, n+k\}$. When $\{\Delta p(t)\}$ is stationary, $Z(t)$ is stationary. Therefore $X(4, t)$ is stationary. Since MACD is its numerator and the denominator is always positive, $X(4, t)$ has the same sign as MACD.

We would like to point out here: we choose $P(t - n - k)$ as the denominator is just for simplicity, which is not the unique choice for the denominator. Any positive linear combination of the near term prices can also be taken as the denominator. The resulted ratio will be always stationary and has the same sign as MACD.

The following is the graph of $X(4, t)$ for every 0.5 second:

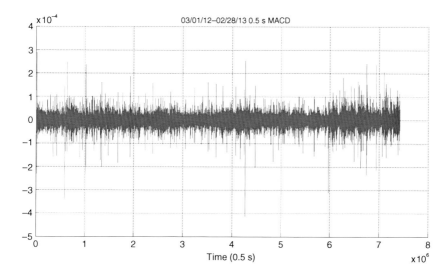

And the following is the graph for every 5 seconds:

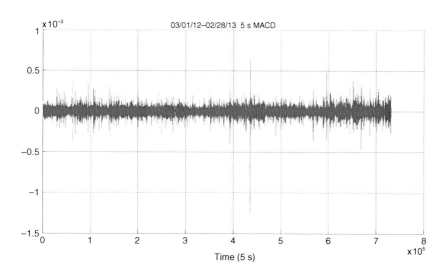

6.6 Rate of Change

The **ROC** is a technical indicator that measures the percentage change between the most recent price and the price "n" periods in the past. It is calculated by using the following formula:

$$ROC_t = \frac{P(t)-P(t-n)}{P(t-n)} = \frac{e^{p(t)}-e^{p(t-n)}}{e^{p(t-n)}} = e^{p(t)-p(t-n)}-1.$$

Usually, ROC appears in percentage form. ROC is classed as a price momentum indicator or a velocity indicator because it measures the ROC or the strength of momentum of change. From the last side, ROC is a function of the logarithm return, so it is stationary. Denote

$$X(5, t) = ROC_t.$$

High ROC is considered as overbought and low ROC is considered as oversold.

The following is the graph of $\{X(5, t)\}$ associated to $P(t)$ which is CSI 300 futures (between March 1, 2012 and February 28, 2013) for every 0.5 second:

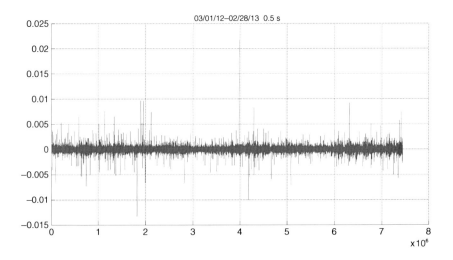

The following is the graph for every 5 seconds:

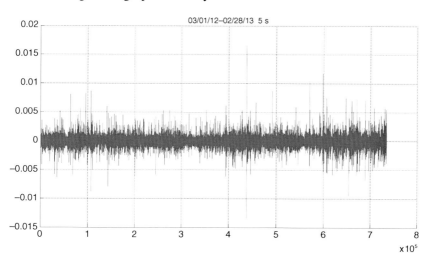

Wen Li discussed $X(5, t)$ for the geometric Brownian motion model in her thesis [16].

6.7 Relative Strength Index

The **RSI** is developed by J. Welles Wilder. RSI is a momentum oscillator that measures the speed and change of price movements. RSI oscillates between 0 and 100. Traditionally, and according to Wilder, RSI is considered overbought when above 70 and oversold when below 30. Signals can also be generated by looking for divergences, failure swings and centerline crossovers. RSI can also be used to identify the general trend.

RSI is an extremely popular momentum indicator that has been featured in a number of articles, interviews and books over the years. In particular, Constance Brown [6] features the concept of bull market and bear market ranges for RSI. Andrew Cardwell introduced positive and negative reversals for RSI. In addition, Cardwell turned the notion of divergence, literally and figuratively, on its head. The formula for RSI is

$$RSI = 100 - \frac{100}{(1 + RS)},$$

where RS = Average of x time units' up closes/Average of x time units' down closes. In mathematics, we denote by I_A the indicator of event A. That is, $I_A = 1$ if A happens, $I_A = 0$ otherwise. Thus,

$$RS = \frac{[P(t) - P(t-1)]I_{[P(t) > P(t-1)]} + \cdots + [P(t-n+1) - P(t-n)]I_{[P(t-n+1) > P(t-n)]}}{[P(t-1) - P(t)](1 - I_{[P(t) > P(t-1)]}) + \cdots + [P(t-n) - P(t-n+1)](1 - I_{[P(t-n+1) > P(t-n)]})}.$$

So we have

$$RSI_t = 100 \frac{RS}{1 + RS},$$

which is a function of the logarithmic returns. Denote $X(6, t) = RSI_t$. Then $X(6, t)$ is stationary.

The following is the graph of $X(6, t)$ for every 0.5 seconds:

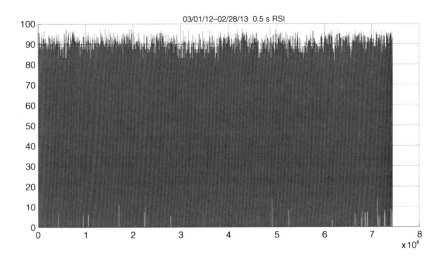

The following is the graph of $X(6, t)$ for every 5 seconds:

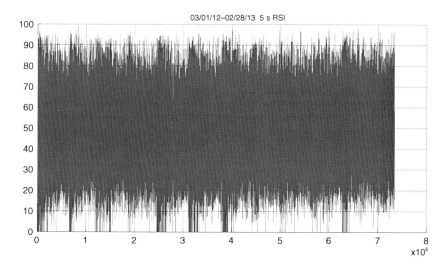

Wei Zhu [33] discussed $X(6, t)$ for geometric Brownian motion model in his thesis.

6.8 Stochastic Oscillators

The **stochastic oscillator** is a momentum indicator that uses support and resistance levels. George Lane promoted this indicator in the 1950s. The Stochastic Oscillator Technical Indicator compares where a security's price closed relative to its price range over a given time period. The stochastic oscillator is displayed as two lines. The main line is called $\%K$. The second line, called $\%D$, is a MA of $\%K$. The $\%K$ line is usually displayed as a solid line and the $\%D$ line is usually displayed as a dotted line.

There are several ways to interpret a stochastic oscillator. Three popular methods include:

- Buy when the oscillator (either $\%K$ or $\%D$) falls below a specific level (e.g., 20) and then rises above that level. Sell when the oscillator rises above a specific level (e.g., 80) and then falls below that level;

- Buy when the %K line rises above the %D line and sell when the %K line falls below the %D line;
- Look for divergences. For instance: where prices are making a series of new highs and the stochastic oscillator is failing to surpass its previous highs.

Denote

$$L(t) = \min\{P(t), P(t-1), \ldots, P(t-n+1)\},$$

as the minimum price of the period of length n, and denote

$$H(t) = \max\{P(t), P(t-1), \ldots, P(t-n+1)\},$$

as the maximum price of that period. Then, we define

$$\%K_t = 100(P(t) - L(t))/(H(t) - L(t)),$$

and define

$$\%D_t = EMA[\%K(t), m],$$

where $m < n$. Usually, we take $m = 3$ and $n = 5, 9,$ or 14. Since

$$\%K_t = 100 \frac{e^{P(t)} - e^{\min\{p(t), p(t-1), \ldots, p(t-n+1)\}}}{e^{\max\{p(t), p(t-1), \ldots, p(t-n+1)\}} - e^{\min\{p(t), p(t-1), \ldots, p(t-n+1)\}}}$$

$$= 100 \frac{e^{p(t)-p(t-n)} - e^{\min\{p(t)-p(t-n), p(t-1)-p(t-n), \ldots, p(t-n+1)-p(t-n)\}}}{e^{\max\{p(t)-p(t-n), p(t-1)-p(t-n), \ldots, p(t-n+1)-p(t-n)\}} - e^{\min\{p(t)-p(t-n), p(t-1)-p(t-n), \ldots, p(t-n+1)-p(t-n)\}}}.$$

When $\{\Delta p(t)\}$ is stationary, so is $\{\%K_t\}$. The following is the graph of $\%K_t$ for every 5 seconds:

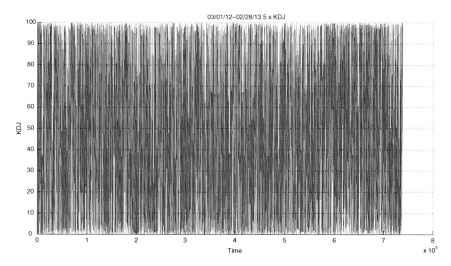

Since $\%D_t$ is the MA of $\%K_t$, $\%D_t$ is also stationary. The following is its graph for every 5 seconds:

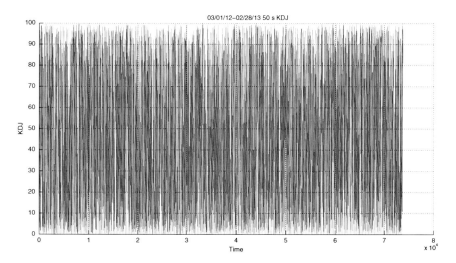

Denote $X(7, t) = \%K_t$, and $X(8, t) = \%D_t$. There are three versions of the stochastic oscillator.

(a) The **fast stochastic oscillator** is based on George Lane's original formulas for $\%K$ and $\%D$ as above. $\%K$ in the fast version that appears rather choppy. $\%D$ is the three-day MA of $\%K$. In fact, Lane used $\%D$ to generate buy or sell signals based on bullish and bearish divergences. Lane asserts that a $\%D$ divergence is the "only signal which will cause you to buy or sell."

(b) The **slow stochastic oscillator** is defined as
Slow $\%K$ = Fast $\%K$ smoothed with three-period MA
Slow $\%D$ = three-period MA of Slow $\%K$.

(c) The **Full Stochastic Oscillator:**
Full $\%K$ = Fast $\%K$ smoothed with X-period MA
Full $\%D$ = X-period MA of Full $\%K$.

Notice that usually $m = 3$, so slow $\%K$ equals to $\%D$ in the fast stochastic oscillator. As $\%D$ in the fast stochastic oscillator is used for signals, the slow stochastic oscillator was introduced to reflect this emphasis.

The stochastic oscillator ranges from 0 to 100. No matter how fast a security advances or declines, the stochastic oscillator will always fluctuate within this range. Traditional settings use 80 as the overbought threshold and 20 as the oversold threshold. These levels can be adjusted to suit analytical needs and security characteristics. Readings above 80 for the 20-time units stochastic oscillator would indicate that the underlying security was trading near the top of its 20-time units high-low range. Readings below 20 occur when a security is trading at the low end of its high-low range.

It is important to note that overbought readings are not necessarily bearish. Securities can become overbought and remain overbought during a strong uptrend. Closing levels that are consistently near the top of the range indicate sustained buying pressure. In a similar vein, oversold readings are not necessarily bullish. Securities can also become oversold and remain oversold during a strong downtrend. Closing levels consistently near the bottom of the range indicate sustained selling pressure. It is, therefore, important to identify the bigger trend and trade in the direction of this trend. Look for occasional oversold readings in an uptrend and ignore frequent overbought readings. Similarly, look for occasional overbought readings in a strong downtrend and ignore frequent oversold readings.

6.9 Directional Movement Index

The **Directional Movement Index (DMI)** was developed by J. Welles Wilder to evaluate the strength of a trend and define periods of sideway trading. The DMI is composed of four curves ($[+DI]$, $[-DI]$, ADX, $ADXR$). The following graph shows the main contract of CSI 300 futures on February 28, 2013 together with *DMI* curves.

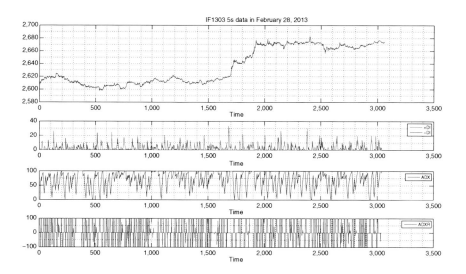

Let us define them in a mathematical way. We select a positive integer k and consider the price change of the period of every k units of time. Denote the high and low in the n-th period by

$$H(n) = \max\{P(nk), P(nk-1), \ldots, P((n-1)k+1)\},$$
$$L(n) = \min\{P(nk), P(nk-1), \ldots, P((n1)k+1)\},$$

and define their increments

$$\Delta H(n) = H(n) - H(n-1) \quad \text{and} \quad \Delta L(n) = L(n-1) - L(n).$$

The DMIs are defined as:

$[+DM] = \Delta H(n)$ (if $\Delta H(n) \geq 0$ and $\Delta H(n) > \Delta L(n)$), or
$[+DM] = 0$ (otherwise);

and

$[-DM] = \Delta L(n)$ (if $\Delta L(n) \geq 0$ and $\Delta H(n) < \Delta L(n)$), or
$[-DM] = 0$ (otherwise).

The **Average Directional Movement (ADM) Indicator** are their EMAs:

$[+ADM] = EMA[+DM, h]$ and $[-ADM] = EMA[-DM, h]$.

The **True Range** (TR) is defined as

$TR(n) = \max\{H(n) - L(n), |H(n) - P(k(n-1))|, |L(n) - P(k(n-1))|\}$.

The **Average True Range** (ATR) of m periods is defined as $MA[TR, m]$. The **Direction Indices** (DIs) are defined by

$[+DI] = 100([+ADM]/ATR)$ and $[-DI] = 100([-ADM]/ATR)$.

The **DMI** is defined by

$DX(n) = 100(|[+ADM] - [-ADM]|)/([+ADM] + [-ADM])$,

and the **Average Direction Movement Index (ADX)** is the EMA of DX. When the $[+DI]$ is above the $[-DI]$, prices are moving up, and ADX measures the strength of the uptrend. When the $[-DMI]$ is above the $[+DMI]$, prices are moving down, and ADX measures the strength of the downtrend.

Finally, define $ADXR$ is the mean of current DX and c period before.

$$ADXR = (DX(n) + DX(n-c))/2.$$

We are going to show all four curves ($[+DI], [-DI], ADX, ADXR$) are stationary processes. Indeed,

$$[+DI] = 100\frac{[+ADM]}{ATR}$$
$$= 100\frac{EMA[[+DM], h]}{MA[TR, m]}$$

$$= 100 \frac{EMA[[+DM], h]}{MA[TR, m]}$$
$$= 100 \frac{EMA[[+DM]/P(nk), h]}{MA[TR/P(nk), m]}. \quad (6.9.1)$$

We are going to show both numerator and denominator of (6.9.1) are stationary. Indeed,

$$\Delta H(n)/P(nk)$$
$$= \max\{1, P(nk-1)/P(nk), \ldots, P(nk-k+1)/P(nk)\}$$
$$- \max\{P((n-1)k)/P(nk), P((n-1)k-1)/P(nk), \ldots,$$
$$\times P((n-1)k-k+1)/P(nk)\}$$
$$= \max\{1, e^{p(n-k-1)-p(nk)}, \ldots, e^{p(nk-k+1)-p(nk)}\}$$
$$- \max\{1, e^{p((n-1)k)-p(nk)}, \ldots, e^{p((n-1)k-k+1)-p(nk)}\},$$

is a function of the logarithmic returns. A similar formula holds for $\Delta L(n)/P(nk)$. Furthermore, we can rewrite

$$[+DM]/P(nk) = \Delta H(n)/P(nk) \text{ (if } \Delta H(n)/P(nk) \geq 0 \text{ and}$$
$$\Delta H(n)/P(nk) > \Delta L(n)/P(nk)),$$
$$\text{or } [+DM]/P(nk) = 0 \quad \text{(otherwise);}$$
$$[-DM]/P(nk) = \Delta L(n)/P(nk) \text{ (if } \Delta L(n)/P(nk) \geq 0 \text{ and}$$
$$\Delta H(n)/P(nk) < \Delta L(n)/P(nk)),$$
$$\text{or } [-DM]/P(nk) = 0 \quad \text{(otherwise).}$$

They are all functions of $(\Delta H(n)/P(nk), \Delta L(n)/P(nk))$. Thus, the numerator of (6.1.9) as their linear combination is also a function of the logarithmic returns. Let us consider the denominator of (6.1.9) now.

$$TR(n)/P(nk) = \max\{H(n)/P(nk) - L(n)/P(nk), |H(n)/P(nk)$$
$$- P((n-1)k)/P(nk)|, |L(n)/P(nk)$$
$$- P((n-1)k)/P(nk)|\},$$

is a function of the logarithmic returns. Therefore, the denominator of (6.1.9) as the *MA* is also a function of the logarithmic return. Thus, the last side of (6.9.1) is a function of $\{\Delta p\}$, which is stationary. By the same method, we can prove that $[-DI]$, ADX and $ADXR$ are all stationary. Now we get four stationary processes:

$$X(9, t) = [+DI]; \ X(10, t) = [-DI]; \ X(11, t) = ADX; \ X(12, t) = ADXR.$$

The following graph is $X(9, t)$ for the main contract of CSI 300 futures (March 1, 2012–February 28, 2012).

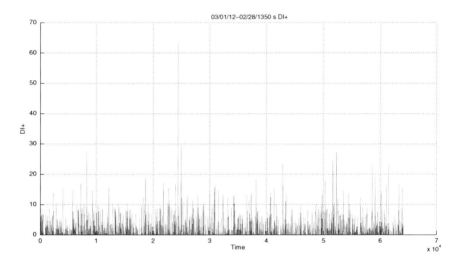

The following is $X(10, t)$ for the same period:

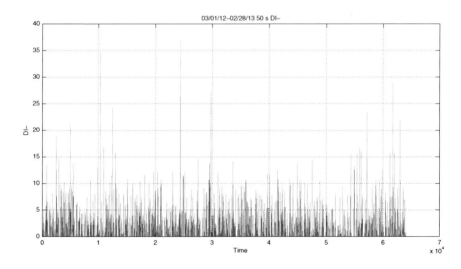

The corresponding $X(11, t)$

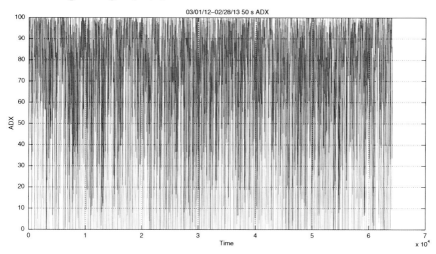

and $X(12, t)$.

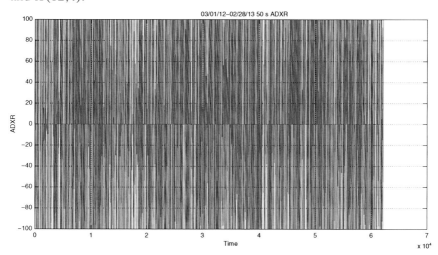

When $[+DI]$ rises above $[-DI]$, it can be considered a signal for an uptrend. When $[+DI]$ crosses below $[-DI]$, it can be considered a signal for a downtrend. According to conventional interpretation, the following criteria should be met for a signal to be considered as valid in most circumstances.

(1) ADX should be rising.
(2) ADX should be above 50.

(3) Confirmation from another indicator is encouraged pointing toward strong trending or volatility characteristics.

A more strict interpretation of the directional moving index calls for a fourth criterion to be met. For an uptrend to be valid, the price of the security must rise above the high of the period that $[+DI]$ up-crossed $[-DI]$. For a downtrend to be valid, the price of the security must dip below the low of the period that $[+DI]$ down-crossed $[-DI]$.

We can list more technical indicators which are connected to the functions of logarithmic returns. For the sake of simplicity of this book, we do not list all of them here and leave them to the readers to identify as exercises. However, there is still one more popular technical indicator which is worthy to discuss. That is Parabolic Sar. We will discuss in Section 6.10.

6.10 Parabolic SAR

Up to now, we get a 12-dimensional function $X(t) = \{X(i, t), i = 1, \ldots 12\}$ of the logarithmic return $\{\Delta p(t)\}$. When $\Delta p(t)$ is **strongly** stationary, so will be $X(t)$. As we know, there are more technical indicators than those 12. As long as they can be written as functions of the logarithmic returns, they will be also strongly stationary and can be added into $X(t)$. If a trader hedges effectively according to some wisely selected $H(t)$ which is a function of $X(t)$, then he may expect stable accumulated profit which will be described in Section 7.1.2.

In the following space, we will discuss another technical indicator which also associated with some stationary process. However, it is not easy to be written as a function of logarithm increments of the price process. Shuai Wang discussed it in his PhD thesis [26].

The **Parabolic Stop and Reverse (Parabolic SAR)** is a method devised by J. Welles Wilder, Jr., to find potential reversals in the market price direction of traded assets. Parabolic SAR is calculated almost independently for each trend in the price. When the price is in an uptrend, the SAR emerges below the price and converges upwards toward it. Similarly, on a downtrend, the SAR emerges above the price and converges downwards. We count every n time units as a "period" (or "step"). For example, when a Chinese trading day is selected as a period, then it contains 4.5 hours or 16,200 seconds. At each step within a trend, the SAR is calculated one period in advance.

That is, next value of SAR is built on data available at current period. The general formula used for this is:

$$SAR_{t+1} = SAR_t + \alpha(EP - SAR_t),$$

where SAR_t and SAR_{t+1} represent the current period and the next period's SAR values, respectively. *EP* (the extreme point) is a record kept during each trend that represents the highest value reached by the price during the current uptrend — or lowest value during a downtrend. During each period, if a new maximum (or minimum) is observed, the *EP* is updated with that value.

α represents the acceleration factor. Usually, this is set initially to a value of 0.02, but can be chosen by the trader. This factor is increased by 0.02 each time a new EP is recorded. What this means is that every time a new EP is observed, it will make the acceleration factor go up. The rate will then quicken to a point where the SAR converges toward the price. To prevent it from getting too large, a maximum value for the acceleration factor is normally set to 0.20. The trader can set these numbers depending on their trading style and the instruments being traded. Generally, it is preferable in stocks trading to set the acceleration factor to 0.01, so that is not too sensitive to local decreases. For commodity or currency trading, the preferred value is 0.02.

The SAR is calculated in this manner for each new period. When the price crosses the SAR value, a new trend direction is then signaled. The SAR must then switch sides. Now let us use mathematical symbol to redefine our Parabolic SAR. In order to get a stationary process, we make a small revision to replace the price $P(t)$ by its logarithm $p(t)$.

Assume that $P(t)$ is in upward case. Let $L'(t)$ be the minimum logarithmic price from the beginning of the trend through the end of t-th period (so there are nt time units) and $H'(t)$ be the maximum logarithmic price for the same periods. Define $S(0) = L'(0)$ and $A(0) = 0.02$.

Then define for the t-th period, the acceleration factor

$$A(t) = \{A(t-1) + 0.02 I_{[H'(t) > H'(t-1)]}\} \wedge 0.2,$$

and SAR

$$S(t) = S(t-1) + A(t-1)[H'(t-1) - S(t-1)].$$

Symmetrically, we define $S(t)$ for the case where $P(t)$ is in a downward trend, which we omitted here.

We take $X(13, t) = p(t) - S(t)$ associated to CSI 300 futures between March 1, 2012 and February 28, 2013. They all past the test:

$H_0 : X(13, t)$ is stationary and $H_1 : X(13, t)$ is not stationary.

The following is the graph of $\{X(13, t)\}$ for every 0.5 second:

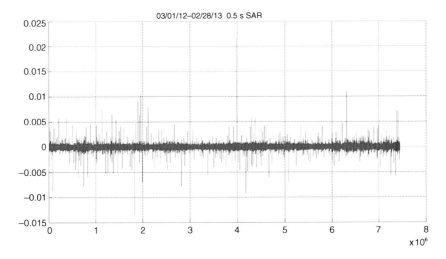

The following is the graph for every 5 seconds:

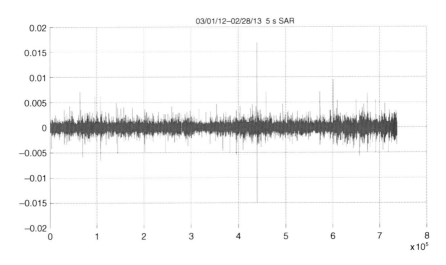

Chapter 7

HFT of a Single Asset

As we have noticed in Section 5.4, the paired trading needs successful execution of four orders for two assets. Therefore, it is natural to ask if a trader can make profit just from repeatedly trading one single financial derivative. There is a big debate on profitability of market timing. We are not interested in taking part in this debate but have a remark as follows. Investors use day or month as the timescale to see the change of the fundamentals, while the time unit is a seconds in high-frequency trading (HFT). When the time is measured by a seconds, the most information one can based on is the momentum shown by technical indicators. From mathematical point of view, the only available information is the stationarity of certain pattern, which can be justified by statistics.

We will only assume that the logarithmic price of some financial instrument has stationary increments, which is a common assumption in most popular price models. In Chapter 6, we showed that some major technical indicators are associated to a multidimensional function $\{X(t)\}$ of the logarithmic returns and $\{(X(t), \Delta p(t))\}$ form a stationary process. We can add more components into that stationary process as there are more technical indicators. If a trader constructs a stationary algorithm according to those technical indicators, then the instant logarithmic returns from each unit of asset form a stationary sequence. Thus, the strong ergodic theorem will assure us a stable mean logarithmic return on each traded unit. That will be our theorem in Section 7.1. We will show two examples of application in Section 7.2.

7.1 Stochastic Integral of Stationary Processes

We say that a function $x(t)$ of time t has **left-limits**, if for each $t > 0$ and any positive sequence $\{s_i\}$ tending to 0, the limit of $\{x(t - s_i)\}$ exists. Denote the left-limit at t as $x(t - 0)$. We say that $x(t)$ has right-limits, if for each $t \geq 0$ and any positive sequence $\{s_i\}$ tending to 0, the limit of $\{x(t + s_i)\}$ exists. Denote the right-limit at t as $x(t + 0)$. If the right-limit (or left-limit) exists and equal to $x(t)$, then we say $x(t)$ is **right-continuous** (left-continuous, respectively) at t.

7.1.1 Ito-Riemann Sums and Their Limit

Let $x(t)$ be a right-continuous function which have left-limits at each $t > 0$. We assume that $h(t)$ is a **right-continuous step-function**, i.e., there exists $0 = t_0 < t_1 < t_2 < \cdots$. such that

$$h(t) = h(t_i) \quad (\text{when } t_i < t < t_{i+1}).$$

For any finite positive t, we can define the integral

$$\int_0^t h(s)dx(s) = \sum_{k=0}^{m-1} h(t_k)[x(t_{k+1}) - x(t_k)] + h(t_m)[x(t) - x(t_m)],$$

where $0 = t_0 < t_1 < t_2 < \cdots < t_m < t \leq t_{m+1}$ exhaust the jumping points of h up to t. This integral is the limit of **Ito-Riemann sum**.

$$\lim_{n \to \infty} \left\{ \sum_{k=0}^{n-1} h\left(\frac{kt}{n}\right) \left[x\left(\frac{(k+1)t}{n}\right) - x\left(\frac{kt}{n}\right) \right] \right\}.$$

Now let $H(t)$ and $X(t)$ be two stochastic process. Suppose their paths are right continuous and have left-limits. If the paths of H are piecewise constants, then the limit

$$\lim_{n \to \infty} \left\{ \sum_{k=0}^{n-1} H\left(\frac{kt}{n}\right) \left[X\left(\frac{(k+1)t}{n}\right) - X\left(\frac{kt}{n}\right) \right] \right\},$$

is called as the **stochastic integral** of H with respect to $X(t)$ and denoted as $\int_0^t H(s)dX(s)$. Here, we should mention that we took the value of H at the beginning of each time-interval $\left(\frac{kT}{N}, \frac{(k+1)T}{N}\right)$ in the above sum, which is different to the Riemann sum of ordinary integral taught in calculus course.

The latter can take any value of H in the interval $\left(\frac{kI}{N}, \frac{(k+1)T}{N}\right)$. The reason to use Ito-Riemann sum instead of Riemann sum is that a trader should pre-determine his hedge before the change of price. The condition imposed on the process $\{H(t)\}$ can be released to more general case when $\{X(t)\}$ is a semi-martingale, for which we still use the same integral notation. However, we will not need it until Section 9.1 where we discuss Black–Scholes model. In the practice of mathematical finance, $\{H(t)\}$ is the hedge of asset and $X(t)$ is the price process. When $P(t)$ is the price of an asset at time t, $H(t)$ is the number of shares held, then

$$\int_0^t H(s)dP(s),$$

is the total gain (or loss) from that stock from the beginning to time t. Since the number of shares one held is always a step function in reality, so our piecewise constant hypothesis on $H(\cdot)$ is realistic. Actually, if $\{P(s)\}$ is the last traded price, then it is also a step function in t. However, mathematically, we need more general assumption on $H(\cdot)$ and $P(\cdot)$ to understand Black–Scholes option theory.

In HFT, one needs to have the total gain $\int_0^t H(s)dP(s)$ growth steadily. We are puzzling if the strong ergodic theorem can help us, as we have shown in Chapter 6 that $d(\log P(t))$ is stationary. That will be the content of Section 7.1.2.

7.1.2 Profit of HFT and Strong Ergodic Theorem

We have discussed the stationarity of the logarithmic return in Section 6.2. Suppose now that $P(t)$ is the last trade price of a financial instrument with stationary logarithmic return Denote $p(t) = \log P(t)$, then $\Delta p(t) = p(t) - p(t - \delta)$ is strongly stationary for any $\delta > 0$. It is a piecewise constant right-continuous function in t.

We consider the following type of hedge: the investor each time can only select between cash or holding the pre-selected asset. We call it a **simple hedge** of that asset. We assume for simplicity that the initial value of a simple hedge is always the initial price $P(0)$ of the asset. There will be no partial selling or partial buying in a simple hedge. That is, if the investor sold his asset at time s and get cash of amount x, then he should use the total amount x to purchase the same asset at a later time t when he wants to buy. Therefore, the digital volume of an asset will appear. In practice, we use the integer part of a digital amount to trade if the error is ignorable. Denote $I(t) = 1$ (if one holds the asset) and $I(t) = 0$ (otherwise). Let us denote by $S(i)$ the starting time of the i-th trading circle and by $T(i)$ the ending time of that circle. That is, one bought at $S(i)$ and sold at $T(i)$. We make the convention that we take the right-continuous version of $\{I(t)\}$. Therefore, $I(t) = 1$ in the random intervals $\{[S(i), T(i)), i = 1, 2, \ldots\}$, and $I(t) = 0$ otherwise.

When N is large enough, one can always assume for simplicity that there is no more than one transaction in each closed time interval $\left[\frac{j}{N}, \frac{j+1}{N}\right]$. If $\left\{\frac{j}{N}\right\} \in \cup_i [S(i), T(i))$, $I\left(\frac{j}{N}\right) = 1$, so

$$\left[P\left(\frac{j+1}{N}\right) \middle/ P\left(\frac{j}{N}\right)\right]^{I(j/N)} = \exp\left\{I\left(\frac{j}{N}\right)\left[p\left(\frac{j+1}{N}\right) - p\left(\frac{j}{N}\right)\right]\right\}.$$

When $\left\{\frac{j}{N}\right\} \in \{\cup_i[S(i), T(i))\}^c$, $I\left(\frac{j}{N}\right) = 0$, the above equality also holds. Therefore, the total realized logarithmic return (without counting transaction costs) by time T is

$$\sum_{j < NT-1} I\left(\frac{j}{N}\right)\left[p\left(\frac{j+1}{N}\right) - p\left(\frac{j}{N}\right)\right]. \qquad (7.1.1)$$

In HFT, the transaction costs form a major factor when one designs a trading strategy. In the China Financial Futures Exchange, the transaction costs of some products are at a small percentage c of total value V of transaction (see Section 2.6). That is, the transaction cost is cV. If one trades the asset of value $P(t)$ at time t, his remaining value is $(1 - c)P(t)$. Thus, the logarithm of remaining value is $\log(1 - c) + p(t)$. Let us deduct the transaction costs from (7.1.1). Since there is no more than one transaction in each closed time interval $\left[\frac{j}{N}, \frac{j+1}{N}\right]$, $\left|I\left(\frac{j}{N}\right) - I\left(\frac{j+1}{N}\right)\right|$ is the number of transaction in that interval. Thus, the total logarithmic affect of transaction costs is

$$\log(1 - c) \sum_{j < NT-1} \left|I\left(\frac{j}{N}\right) - I\left(\frac{j+1}{N}\right)\right|. \qquad (7.1.2)$$

For simplicity, we assume that there is no trade at time 0. Combining (7.1.1) and (7.1.2), the total realized logarithmic return by time T is

$$\sum_{j<NT-1} \left\{ I\left(\frac{j}{N}\right)\left[p\left(\frac{j+1}{N}\right) - p\left(\frac{j}{N}\right)\right] \right.$$
$$\left. + \log(1-c)\left|I\left(\frac{j}{N}\right) - I\left(\frac{j+1}{N}\right)\right| \right\}. \quad (7.1.3)$$

When $N \to \infty$, the above sum converges to

$$\int_0^T I(s)dp(s) + \log(1-c)K(T)$$
$$= \sum_{\{i;T(i)<T\}} [p(T(i)) - p(S(i))] + \log(1-c)K(T), \quad (7.1.4)$$

where $K(T)$ is the number of transactions before T. Here, we count buying and selling as two transactions.

When $\{(I(.), \Delta p(.))\}$ is stationary, so is each term in the sum of (7.1.3). Thus, the mean logarithmic return is

$$\frac{1}{T} \sum_{\{i;T(i)<T\}} [p(T(i)) - p(S(i)) + 2\log(1-c)]. \quad (7.1.5)$$

Therefore, we can apply the strong ergodic theorem to deduce that the last side of (7.1.5) converges.

From the above discussion, the total gain up to T is given by

$$P(0) \left\{ \exp\left\{ \int_0^T I(s)dp(s) + \log(1-c)K(T) \right\} - 1 \right\}$$
$$= P(0) \left\{ \exp\left\{ \sum_{\{i;T(i)<T\}} [p(T(i)) - p(S(i)) + 2\log(1-c)] \right\} - 1 \right\}.$$

Now let us consider the general situation. Denote by $H(t)$ the number of simple hedges at time t. $H(t)$ is always bounded. When $H(t) > 1$, $H(j/N)[p((j+1)/N) - p(j/N)]$ has no financial meaning, so we need to introduce a new method. We introduce several processes $\{I(j,t)\}$ which are functions of $\{H(t)\}$ as follows:

$$I(j,t) = 1 \text{ (if } j \le H(t)) \quad \text{and} \quad I(j,t) = 0 \text{ (if } j > H(t)). \quad (7.1.6)$$

Then, $\sum_j I(j, t) = H(t)$. For the convenience of our non-mathematician readers, we give the following table to show the relation between $H(t)$ and $\{I(j, t)\}$. We assume that $H(t) \leq 7$ for simplicity.

$H(t)$	$I(1, t)$	$I(2, t)$	$I(3, t)$	$I(4, t)$	$I(5, t)$	$I(6, t)$	$I(7, 5)$
1	1	0	0	0	0	0	0
2	1	1	0	0	0	0	0
3	1	1	1	0	0	0	0
4	1	1	1	1	0	0	0
5	1	1	1	1	1	0	0
6	1	1	1	1	1	1	0
7	1	1	1	1	1	1	1

From previous discussion for the case where $H(t) \leq 1$, we deduce that

$$P(0)\left\{\exp\left\{\int_0^T I(j, s)dp(s) + \log(1 - c)K(j, T)\right\} - 1\right\},$$

will be the gain by time T from hedging $I(j, s)$, where $K(j, T)$ is the total number of trading according to $\{I(j, s)\}$. Therefore, the total gain from hedging $\{H(t)\}$ will be the sum

$$P(0) \sum_j \left\{\exp\left\{\int_0^T I(j, s)dp(s) + \log(1 - c)K(j, T)\right\} - 1\right\}. \quad (7.1.7)$$

Since each term in the sum has stably growth, so will be their sum. We write this computation as the following:

Theorem Let $H(t)$ be the number of simple hedges of an asset at time t and $I(.,.)$ be given by (7.1.6). Denote by $K(j, T)$ the number of transactions (assuming $K(j, 0) = 0$) corresponding to $\{I(j, s)\}$ before T and denote the transaction costs per trade as $cP(t)$. If $\{(H(t), \Delta p(.))\}$ is stationary, then the total net profit by hedge $\{H(t)\}$ is given by (7.1.7)

Moreover, the mean logarithmic return of each term in (7.1.7)

$$\frac{\left\{\int_0^T I(j, s)dp(s) - \log(1 + c)K(j, T)\right\}}{T}$$

converge when $T \to \infty$ for each $j = 1, 2, \ldots$

Some pure mathematicians may worry about how large T should be in order to get the above convergence? When our time unit is 0.5 second, four trading hours will give $T = 28,800$. So the terminal time T does not need to be a few months. We also would like to mention that in HFT practice, the strategies and algorithms depend on the competitions. When the competitors change, the algorithm should be changed as well. One should not think there is some constant algorithm which can work for years. Finally, for the reason of risk control, we should always set the upper bound for $\{H(t)\}$ so that the possible loss $H(t)[P(t+a) - P(t)]$ is affordable even in extreme cases.

In Chapter 6, we showed that the main technical indicators are functions of $\{\Delta p(.)\}$. Thus, if one hedges according to the main technical indicators, $\{(H(t), \Delta p(.))\}$ satisfy the condition of the above theorem.

Remark We have to emphasize here: we did not include the time length for the order to be executed by the exchange in the above theorem and thus we may not be able to get our orders executed exactly according to $\{I(j, s)\}$. A wise design of strategy should have enough stability with respect to small perturbation of $\{I(j, s)\}$. For the same reason, a testing trade in the real market is necessary except for very experienced trader, who can estimate the bound of errors.

7.1.3 Sharpe Ratio Test for HFT

In 1966, William Forsyth Sharpe developed what is now known as the Sharpe ratio. Sharpe originally called it the "reward-to-variability" ratio before it began being called the Sharpe ratio by later academics and financial operators. The definition was:

$$S = E[R - R_f]/(\text{Var}(R))^{1/2},$$

where R is the return of the portfolio, R_f is the return of the reference asset, $E[R - R_f]$ is their expected difference and $(\text{Var}(R))^{1/2}$ is the standard deviation of R. Sharpe's 1994 revision acknowledged that the basis of comparison should be an applicable benchmark, which changes with time. After this revision, the definition is:

$$S = E[R - R_b]/(\text{Var}(R - R_b))^{1/2}.$$

When R_b is a constant risk-free return throughout the period, $E[R_b] = R_b$ and $\text{Var}(R - R_b) = \text{Var}(R)$. Since Chinese Government Bond yields are higher than the Western countries', we use often the Government Bond yields for R_b in Chinese market. Thus, we may use

$$S = (E[R] - R_b)/(\text{Var}(R))^{1/2}, \qquad (7.1.8)$$

for our *ex-ante* Sharpe ratio. Now new problem is raised: how to calculate $E[R]$ and $\text{Var}(R)$? Several statistical tests of the Sharpe ratio have been proposed. These include those proposed by Jobson and Korkie, and Gibbons, Ross and Shanken. For HFT, we have a new method to estimate Sharpe ratio based on the strong ergodic theorem. When we have a simple hedge, then from (7.1.3), the stationarity of

$$R = I\left(\frac{j}{N}\right)\left[p\left(\frac{j+1}{N}\right) - p\left(\frac{j}{N}\right)\right] + \log(1-c)\left|I\left(\frac{j}{N}\right) - I\left(\frac{j+1}{N}\right)\right|,$$

follows from that of $\{(I, \Delta p)\}$. Thus, according to the strong ergodic theorem,

$$E[R] \approx \frac{1}{T} \sum_{\{i; T(i) < T\}} \left\{ I\left(\frac{j}{N}\right)\left[p\left(\frac{j+1}{N}\right) - p\left(\frac{j}{N}\right)\right] + \log(1-c)\left|I\left(\frac{j}{N}\right) - I\left(\frac{j+1}{N}\right)\right|\right\},$$

and

$$E[R^2] \approx \frac{1}{T} \sum_{\{i; T(i) < T\}} \left\{ I\left(\frac{j}{N}\right)\left[p\left(\frac{j+1}{N}\right) - p\left(\frac{j}{N}\right)\right] + \log(1-c)\left|I\left(\frac{j}{N}\right) - I\left(\frac{j+1}{N}\right)\right|\right\}^2.$$

Thus,

$$\begin{aligned}\text{Var}(R) &\approx E[R^2] + E^2[R] \\ &= \frac{1}{T} \sum_{\{i; T(i) < T\}} \left\{ I\left(\frac{j}{N}\right)\left[p\left(\frac{j+1}{N}\right) - p\left(\frac{j}{N}\right)\right] \right. \\ &\quad \left. + \log(1-c)\left|I\left(\frac{j}{N}\right) - I\left(\frac{j+1}{N}\right)\right|\right\}^2 \\ &\quad - \left(\frac{1}{T} \sum_{\{i; T(i) < T\}} \left\{ I\left(\frac{j}{N}\right)\left[p\left(\frac{j+1}{N}\right) - p\left(\frac{j}{N}\right)\right] \right.\right. \\ &\quad \left.\left. + \log(1-c)\left|I\left(\frac{j}{N}\right) - I\left(\frac{j+1}{N}\right)\right|\right\}\right)^2.\end{aligned}$$

Therefore, based on the historical data, we can easily get a theoretical Sharpe ratio for simple hedge HFT from (7.1.8). The above method can be easily extended to the more general cases.

However, a same remark as in Section 7.1.2 should be still added. The above Sharpe ratio estimate did not include the error term caused by the execution time of the exchange.

7.2 Two Examples

Our both examples show the application of Moving Average Convergence–Divergence (MACD). However, the techniques applied are slightly different. We should emphasize here that, the slight difference of techniques might cause crucial difference in HFT, as the profit is so low in each trade.

7.2.1 The First Example

The first example is from Yu Zhou's computation (see [4]). Here is the 50-second chart of CSI 300 futures main contract IF1204 on March 16, 2012:

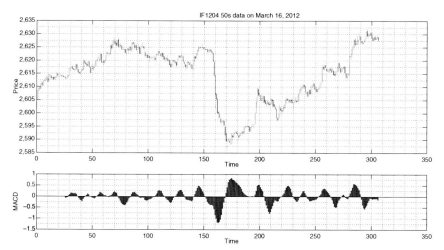

We consider the graph of the revised MACD $X(4, t)$ introduced in Section 6.5. It is easy to see that its graph looks like a strip. We set $a = -0.00019$ as the lower bound and $b = 0.00023$ as the upper bound (the horizontal lines).

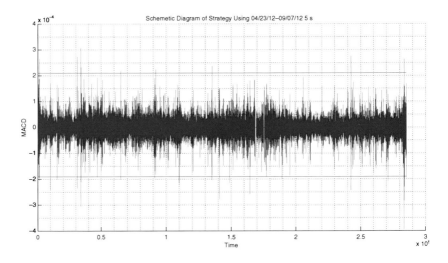

The algorithm is

(1) to buy when $X(4, t)$ reaches the lower bound a from below and then to sell when $X(4, t)$ reaches either 0 or $1.2a$ (whichever arrives first).
(2) to sell when $X(4, t)$ reaches the upper bound b from above and then to buy back when $X(4, t)$ reaches either 0 or $1.2b$ (whichever arrives first).

The accumulated profit from April 23, 2012 through September 7, 2012 is shown in the following graph:

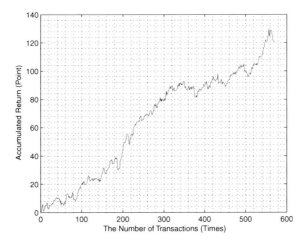

We can see that, between the 350th and the 500th trades, the profit was flat. In the real market, a strategy should be adjusted at least weekly to meet competition. Our example here is just to show the existence of statistical arbitrage. The next example is based on a smoothed MACD designed by Chang Liu [17, 18].

7.2.2 The Second Example

In the previous example, one can see from the graph of accumulated profit that the profit of using MACD shrunk after the summer of 2012. That may be the result of competition. MACD is a very popular technical indicator, so many traders may use it. The only difference is possibly the timescales. In order to get a better technical indicator above the competitors, we introduce the second example from [18] which is a part of Chang Liu's PhD thesis [17]. Instead of using directly the MACD indicator, Liu suggested **Smoothed MACD** as the following: Denote for $q < p < m < n$ (usually take $(q, p, m, n) = (3, 9, 12, 26)$),

$$M(t) = \underline{EMA}\,[\underline{EMA}[P, m] - \underline{EMA}[P, n], q],$$

and

$$S(t) = \underline{EMA}\,[\underline{EMA}[P, m] - \underline{EMA}[P, n], p].$$

Let $B(t) = \underline{EMA}[M-S, q](t)$ be their smoothed difference Liu used (M, S, B) to form the three indicators of smoothed MACD. For different timescales, we have different MACDs. So we use the notation $(M^{(i)}(t), S^{(i)}(t), B^{(i)}(t))$ to denote the corresponding terms when the smoothed MACD indicator is corresponding to the timescale of i seconds.

The following is the graph of CSI 300 futures main contract on January 6, 2012, the corresponding MACD for $i = 10$ and $i = 60$ seconds, and the return points on that day.

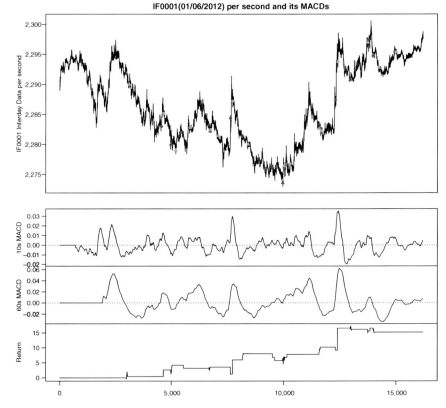

Liu do not use $X(4, t)$ directly but use $\{M^{(i)}(t), S^{(i)}(t), B^{(i)}(t)\}$ with different timescales instead. After replacing EMAs by EMAs (which will not create big differences in calculation but the speed), those indicators are linear combinations of the prices at different time. We can still transform them into stationary ones after dividing them by current price as shown in Chapter 6.

The following is Liu's strategy:

(1) **Buy open and sell close:** When $M^{(60)}(t) > 0$ and $B^{(60)}(t) > 0$, $B^{(20)}(t)$ is turning from negative to positive; that is a "buy" signal. After that, one closes his position when one of the following cases happens:

   (a) $M^{(60)}(t)$ is up but $M^{(20)}(t)$ is down;
   (b) All $M^{(60)}(t)$, $M^{(20)}(t)$ and $M^{(6)}(t)$ are down;

(c) Both $M^{(60)}(t)$ and $M^{(6)}(t)$ are down but $M^{(20)}(t)$ is up, if one has positive profit.

(2) **Short open and buy close:** When $M^{(60)}(t) < 0$ and $B^{(60)}(t) < 0$, $B^{(20)}(t)$ is turning from positive to negative; that is a "sell" signal. After that, one closes his position when one of the following cases happens:

(a) $M^{(60)}(t)$ is down but $M^{(20)}(t)$ is up;
(b) All $M^{(60)}(t)$, $M^{(20)}(t)$ and $M^{(6)}(t)$ are up;
(c) Both $M^{(60)}(t)$ and $M^{(6)}(t)$ are up but $M^{(20)}(t)$ is down, if one has positive profit.

If we just bid at quoted sell price and sell at quoted bid price in above two cases, the annualized yield is 94% and Sharp ratio is around 4.75 from this strategy. The following graph shows the calculated accumulated daily profits in 2012 (after deduction of transaction costs):

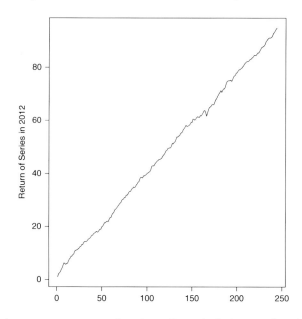

Since the average return of each trading circle is very low in HFT, it is natural to think if we can give counter offer to raise the returns? We will discuss this issue in Chapter 8.

Chapter 8

Bid, Ask and Trade Prices

The bid-ask spread is the difference between the prices quoted for an immediate purchase (ask) and an immediate sell (bid). In Section 3.2.2, we have a few graphs to show the relative position of bid-ask-last price. For the statistical arbitrage using paired products, one needs to buy at the ask price and to sell at the bid price to increase the rate of successes. However, the profit of each trade is so thin in high-frequency trading (HFT). When one trade Chinese stocks index futures contracts, one tick (0.2 point) of difference will be roughly the profit of each trade. Thus, it is natural to ask if there is other way to give the trading order. That is an important problem in HFT.

If we look closely the graphs shown in Section 3.2.2, a big part of the trade prices was neither bid nor ask prices quoted at the previous moment. Is there any statistical way to treat this problem? Our problem here is how to use the bid-ask spread analysis to discover statistical arbitrage. We discussed the rates of success when one inserts trade orders around the bid-ask prices [18]. The following table shows how the rate of success depends on the orders and execution time to trade CSI 300 futures in 2012. For example, if one gives buy order one tick below (i.e., -0.2) the ask price, then the rate of success within 1 second is 82.24%, and within 5 seconds is 93%.

Chang Liu studied in details this problem in his PhD thesis [17]. There have been a lot of studies on the relation between the trading volumes and trading ranges. It is well-known that the trading range is roughly proportional to the trading volume. And there have also been many discussions on the prediction of the trading volumes. However, in HFT, the

Rate of Success with 1–5 seconds

Order	Type	1 s (%)	2 s (%)	3 s (%)	4 s (%)	5 s (%)
Bid − 0.2	Sell	99.8	99.91	99.93	99.94	99.95
Ask + 0.2	Buy	99.82	99.92	99.94	99.95	99.95
Bid	Sell	99.44	99.73	99.8	99.83	99.84
Ask	Buy	99.46	99.74	99.81	99.83	99.85
Bid + 0.2	Sell	81.82	88.3	90.78	91.83	92.62
Ask − 0.2	Buy	82.24	88.73	90.92	92.25	93
Bid + 0.4	Sell	52.46	62.16	67.54	70.75	73.39
Ask − 0.4	Buy	52.63	62.61	68.03	71.24	74.07

traders are more interested in finding profitable strategies based on those theoretical results. In his thesis [17], C. Liu revealed one of such attempts.

The following graph shows the relation between the trading range and volume of CSI 300 futures between 9:30 and 9:55, on January 6, 2012.

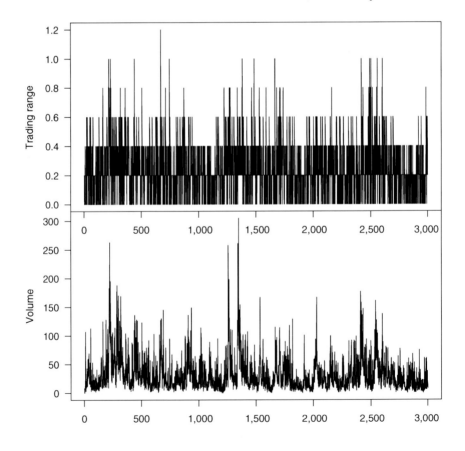

Therefore, in order to successfully buy lower and sell higher than the corresponding quoted price, one has to be able to predict the trading volume for the next moment. Chang Liu used a formula for the prediction and obtained a result shown in the following graph, which shows the real volume compared with the predicted one (0.5 second in advance) of CSI 300 futures between 9:30 and 9:55, on January 6, 2012.

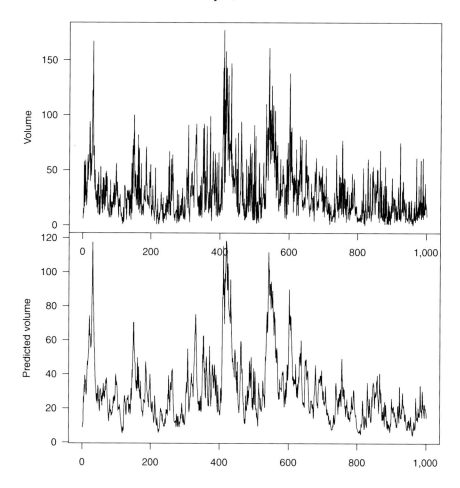

Chang Liu showed also in his thesis that [17], if one inserts buy order at one tick less than the quoted ask price and inserts sell order at one tick higher than the quoted bid price when the predicted volume arises, then the entire profit of the second example shown in Section 7.2.2 will be higher.

The Sharp ratio will be raised from the original 4.75 to 7.14 and annualized yield will be raised to 127.7% from 94% (see the following bolder line vs. the original one).

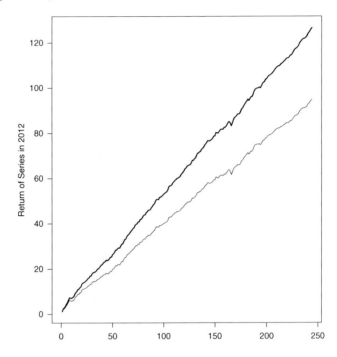

Since this is related to a subject we are still making progress at our Seminars in Shanghai, we omit some technical details here. Interested readers may refer to Ref. [18] for partial results.

Chapter 9

Financial Engineering

The **Financial Engineering** (FE) is a multidisciplinary field involving financial theory, the methods of engineering, the tools of mathematics and the practice of programming. It has also been defined as the application of technical methods, especially from mathematical finance and computational finance, in the practice of finance. A lot of high technology has been applied in FE, which is why one calls it as engineering. FE is quite different from traditional financial study. In broadest definition, anyone who uses technical tools in finance could be called a financial engineer, for example, any computer programmer in a bank or any statistician in a government economic bureau. However, most practitioners restrict the term to someone educated in the full range of tools of modern finance and whose work is informed by financial theory. It is sometimes restricted even further, to cover only those originating new financial products and strategies.

In FE, the most difficult part which needs modern high technology is program trading. Certainly, at least seems to us, the most difficult part of program trading is HFT, which needs even the competition of speed. In this chapter, we will briefly review some basic elements of mathematical finance, statistical finance, behavioral finance and computational finance.

9.1 Mathematical Finance

Mathematical finance is a field of applied mathematics, concerned with financial markets. Some mathematicians also call it as **Financial Mathematics** to emphasize their main interests is to find mathematical problems originated from finance to develop mathematics. In mathematical finance, mathematicians intend to establish rigorous mathematical models to understand the financial market. It has been critically remarked often that (see various editions of Wikipedia "Mathematical Finance"), some financial mathematicians might derive and extend the mathematical or numerical models without necessarily establishing a close link to financial theory and taking observed market prices as input.

We discussed arbitrage in Section 3.2. The fundamental theorem of arbitrage-free pricing is one of the key theorems in mathematical finance, while the Black–Scholes equation and formula are amongst the key results, which will be explained later. Mathematically, the **arbitrage-free hypothesis** is assuming that there is no portfolio with value $\{V(t)\}$ and a **bounded** time T such that:

(i) $V(0) = 0$;
(ii) $P[V(T) \geq 0] = 1$;
(iii) $P[V(T) > 0] > 0$.

We would like to make a remark here. The above definition of arbitrage-free hypothesis is not contradicting the statistical arbitrages discussed in the previous chapters, as the latter needs T being unbounded to satisfy Condition (ii). Let us consider an example of statistical arbitrage but no bounded time to claim the victory with 100% of probability as follows. Two gamblers A and B roll a fair dies continuously. Gambler A always bet one dollar on getting "6", and Gambler B always bet one dollar for "not 6". It is easy to see that the expected gain for Gambler B is 4/6. The law of large number tells us that Gambler B's average gain per time is approximate 4/6 as well. However, there is still a probability $(1/6)^N$ that Gambler A is always a winner for all first N rolling. Therefore, there is no arbitrage satisfying Condition (ii).

Black–Scholes assumed the geometric Brownian motion as the price model of a stock share:

$$P(t) = P(0) \exp\{\sigma W(t) + \mu t\}.$$

Denote by T the mature time of a call option for buying a share at price K. Then, the gain at time T will be $(P(T) - k)^+$, which is the positive part of $(P(T) - k)$. That is, $(P(T) - k)^+ = (P(T) - k)$ when $(P(T) - k)$ is positive and $(P(T) - k)^+ = 0$ otherwise. The Black–Scholes theory points out that, if the price $P(t)$ is a geometric Brownian motion, then there are two continuous process $H(t)$ and $V(t, P(t))$ such that the stochastic integral holds:

$$(P(T) - k)^+ = V(t, P(t)) + \int_t^T H(s) dP(s),$$

where we assume interests free and no transaction costs for simplification. If an option seller hedges the number of shares according to $\{H(s)\}$, then in the time interval $[t, T]$, he has the gain $(P(T) - k)^+ - V(t, P(t))$. Therefore, $V(t, P(t))$ is the fair price of this option at time t. A stunning implication of Black–Scholes fundamental theorem of arbitrage-free is that $V(t, P(t))$ is just equal to $E[(P(T) - k)^+]$ when one assumes $\mu = -\sigma^2/2$. σ is the so-called **volatility**. One can estimate σ by the standard deviation of $p(t) = \log P(t)$ when m is enough large:

$$\sigma^2 \approx \sum_{k=0}^{m-1} [p((k+1)/m) - p(k/m)]^2.$$

One can easily see that $\sigma W(t)$ is a Gaussian distributed random variable with mean 0 and variance $t\sigma^2$, so we get the **Black–Scholes formula** from taking the conditional expectation:

$$V(t, x) = E[(P(T-t) - k)^+ | P(t) = x]$$
$$= \int \frac{1}{\sigma \sqrt{2t\pi}} (x \exp\{y - ((T-t)\sigma^2/2)\} - k)^+$$
$$\times \exp\{-y^2/2(T-t)\sigma^2\} dy.$$

$V(t, x)$ satisfies a partial differential equation which is called **Black–Scholes equation**:

$$V_t + \frac{1}{2}\sigma^2 x^2 V_{xx} = 0.$$

In the above arguments, mathematicians ignored the transaction costs. There have been also quite a few literatures which discussed the transaction costs. It is also known that there is a quite big gap between the theoretical price $V(t, P(t))$ and the real market option price. In order to get better approximation, many mathematicians developed much more sophisticated models. We will not discuss this issue here. However, we would like to point out: the price process $P(t)$ here is the (last) trading price. From our discussion in Chapter 8, one can easily find that the accumulated transaction costs should include the accumulated bid-ask gap, which is actually the source of the profit of HFT.

Mathematical finance also overlaps heavily with the field of computational finance. The latter focuses on application, while the former focuses on modeling and derivation, often by help of stochastic asset models. In general, there exist two separate branches of finance that require advanced quantitative techniques: derivatives pricing on the one hand, and risk- and portfolio management on the other hand.

9.2 Statistical Finance

Statistics is closely related to finance from the beginning of that the investors wanted to study the financial phenomena quantitatively. However, the modern **Statistical finance** can be considered more or less as the application of econophysics to financial markets. Econophysics was started in the mid-1990s by several physicists working in the subfield of statistical mechanics. One driving force behind econophysics arising at that time was the sudden availability of large amounts of financial data, starting in the 1980s. The traditional explanations and approaches of economists usually prioritized simplified approaches for the sake of soluble theoretical models over agreement with empirical data. Those physicists applied tools and methods from physics, first to try to match financial data sets, and then to explain more general economic phenomena. It became apparent that traditional methods of analysis were insufficient. The standard economic methods dealt with homogeneous agents and equilibrium, while many of the more interesting phenomena in financial markets fundamentally depended on heterogeneous agents and far-from-equilibrium situations.

There are three main themes in statistical finance:

(a) Empirical studies and the discovery of interesting universal features of financial time series;
(b) Use of empirical results to design better models of risk and derivative pricing;
(c) Study of "agent-based models" in order to unveil the basic mechanisms that are responsible for the statistical "anomalies" observed in financial time series.

Many studies from statistical finance are useful for FE. A financial engineer can use the results from financial statisticians to create the algorithm of trade. We have shown an example of such application in the last section of Chapter 5.

Possibly, one of the major differences between the statistical finance and FE is that the latter is more interested in finding the strategies and algorithms of making profits from historical data. The three basic steps for establishing a HFT strategy are:

(a) Based on historical data to construct a profitable stationary strategy;
(b) Simulate the transactions based on the new strategy to verify the effectiveness of the new strategy;
(c) Trade on real market and revise the algorithm according to its feedback from the real trading data.

As we have seen from the previous chapters, a high-frequency trader may gain profit based on the strong ergodic theorem of stationary time series. And the stationarity time series is one of the major research interests of modern mathematical statistics.

9.3 Behavioral Finance

Behavioral finance studies the effects of social, cognitive and emotional factors on the economic decisions and the consequences for market prices, returns and the resource allocation. The fields are primarily concerned with the bounds of rationality of economic agents. Behavioral models typically integrate insights from psychology with neo-classical economic theory; in so doing, these behavioral models cover a range of concepts, methods and fields.

The study of behavioral economics includes how market decisions are made and the mechanisms that drive public choice, such as biases toward promoting self-interest.

There are three prevalent themes in behavioral finances:

- **Heuristics**
  People often make decisions based on approximate rules of thumb and not strict logic.
- **Framing**
  The collection of anecdotes and stereotypes that make up the mental emotional filters individuals rely on to understand and respond to events.

- **Market inefficiencies**
  These include mis-pricings and non-rational decision making.

A typical example of quantitative analysis of behavioral finance is the well-known Hong–Stein's model [13]. Hong and Stein have separated the traders into two groups: the momentum traders and the news-watchers. News-watchers affected the price gradually (according to the speed of leaking of news?) and momentum traders affected the price according to the increments of the latter. It seems to us that, technical analysis reflects the activities of the momentum traders, and some sudden changes of price reflect the activity of news watchers. Hong and Stein's model is given by the following formula:

$$P(t) = D_t + \frac{z-1}{z}\varepsilon_{t+1} + \frac{z-2}{z}\varepsilon_{t+2} + \cdots + \frac{1}{z}\varepsilon_{t+z-1}$$
$$- Q + jA + \sum_{i=1}^{j} \phi \Delta P(t-i),$$

where $\{\varepsilon_i\}$ are independent identically distributed Gaussian random variables and $D_t = D_0 + \sum_{i=1}^{t} \varepsilon_i$. Since Hong–Stein's model just reflected some bare market behaviors, we can only consider it as an inspiring heuristic model. The further studies are attractive. Shuai Wang discussed this model in his thesis [27] (see also Ref. [26]). In HFT, the time unit is not more than a second; therefore, the news should arrive at the scale of every k seconds for a very large k. Thus, the influence of news is not so frequent and the momentum plays a more important role. That may be an explanation from the behavioral finance to the efficiency of technical analysis in HFT.

9.4 Computational Finance

Computational finance is a branch of applied computer science that deals with problems of practical interests in finance. Some slightly different definitions are the study of data and algorithms currently used in finance, and the mathematics of computer programs that realize financial models or systems.

Computational finance emphasizes practical numerical methods rather than mathematical proofs and focuses on techniques that apply directly to economic analysis. It is an interdisciplinary field between mathematical finance and numerical methods. Two major areas in computational finance are the following:

(a) efficient and accurate computation of fair values of financial securities;
(b) modeling of stochastic price series.

It seems to us that there is a third area which is much more challenging. That is to optimize the speed of computer hardware and software, the internet connection of the systems, ... all the problems discussed in Chapter 4. As an example, the definition of EMA being better than EMA (see Section 6.3) on fast computation was not known by most financial mathematicians. However, such a replacement is crucial for the speed in HFT.

Nowadays, most of stock trading is done through massively and globally interlinked computer systems. The rates of these transactions are now limited only by technology and progressively by the speed of light. So a costly arms race has begun for telecommunications and network links that can give traders a competitive edge as small as a few tens of microseconds. Typically, a HFT firm buys and sells financial instruments while holding on to them for perhaps just fractions of a second. High-frequency traders make money by exploiting tiny and fleeting disequilibrium in the markets.

As we showed in Chapter 4, one reason that the high-frequency traders can beat others to the punch is that they often locate their computers in data centers run by the exchanges. It can command premium prices for that space because closer physical proximity means faster access to the exchange. Though light speed is 300 million meter per second, it is just 300 meter per microsecond. High-frequency traders should always try to get the best place in computer rooms so that their trading orders can be executed earlier than the others.

HFT firms also gain an edge by having the fastest telecommunications link possible between distant trading centers where the prices of what is being bought and sold are fundamentally related. The fate of some companies whose stocks are being traded in New York City, for example, hinges on the price of commodities being traded in Chicago, and vice versa. If the computerized trading platform in one of these cities has access to information about the market in the other one sooner than anyone else — even just a few milliseconds sooner — it can execute profitable trades.

Big data is the term for a collection of data sets so large and complex that it becomes difficult to process using on-hand database management tools or traditional data processing applications. The challenges include capture, curation, storage, search, sharing, transfer, analysis and visualization. Our book shows only some algorithms of HFT which are not too far from traditional data processing and human eyes. More advanced algorithms can evaluate thousands of securities with complex mathematical tools, far beyond human capacity. Algorithms also can combine and analyze data to reveal insights not readily apparent to the human eyes. This is where true innovation can happen — there is a seemingly endless amount of data available to us today, and with the right tools, financial modeling becomes limited only by the brain power and imagination of the quant at work. The future of finance is algorithms looking beyond the order book to the rapidly expanding universe of time-series data, opening the market to the power of quant creativity.

Chapter 10

Debate and Future

If you ever searched the phrase "day trader" online before this century, you would be told that more than 99% of day traders lost their money finally. This was possibly true at that time and is possibly partially true nowadays. The advance of science and technology changed all, including the definition of "day trader". High-frequency trading (HFT) is one of those day trading activities.

Through the discussion of the previous chapters, it is easy to see that the speed and strategy are two key factors for HFT. In today's highly competitive world of algorithmic trading, it is essential for HFT firms to host their key trading machine as close as possible to exchange's matching engines in order to remain competitive. Colocation (also spelled colocation, collocation) hosting means that the exchange participants trading applications are located in the same data center as Exchange's matching engines. A colocation center is a type of data center where equipments, spaces and bandwidths are available for rental to individual customers. Colocation facilities of exchanges provide space, power, cooling and physical security for the server, storage and networking equipment of other firms — and connect them to a variety of telecommunications and network service providers — with a minimum of cost and complexity. Therefore, HFT has a possibly too high minimum entrance cost to a retail investor. The best way for a retail investor to participate in HFT is possibly through a HFT firm. That is just like the safest way of investing in stock market is to purchase mutual funds unless one thinks his own knowledge and technique of timing can beat a group of professional investors who spend time to do fundamental analysis, to understand financial reports, have a quick access to exchange, and so on.

Program trading is now a popular practice in the US market. For example, when a retail investor trades stock options, he can easily find that the bid-ask spreads exceed 10% of the near-term option prices and the option prices follow very closely the price change of the corresponding stock share. It is not difficult to discover that retail investors are actually trading against the computer. From the discussions of the previous chapters, it is easy to see that HFT is the top technique skill for program trading.

We have been asked frequently what the impacts of HFT are. We attempt to answer as follows:

(a) **Liquidity** is the first impact. As HFT is playing a very large trading volume in the market, the total trading volume has been increased a lot. Traditional view of liquidity based on trading volume is not so accurate now. The recent review of liquidity is based on bid-ask spread, as a narrow bid-ask spread means everybody can buy or sell at a low cost, which benefit both large companies and individual investors. Introduction of HFT has obviously narrowed this spread.

(b) **Volatility** is also an important impact. Somebody believes that HFT traders may detect the recent market trend and push it before other traditional traders in a very large volume, which cause a larger volatility. The problem is if any HFT trader pushes the market too much, the market will soon get back and that trader will be punished by the market. Therefore no reasonable high-frequency trader will do it unless he is not aware of it, which exists also among traditional traders. Thus, volatility is not seemed to be changed much by HFT. Narrowed bid-ask spread by HFT may even help to reduce the volatility.

(c) The third impact is on short-term **price discovery**. As HFT traders are using sophisticated models which should be believed useful for price discovery. However, high-frequency traders normally only focus on short-term direction instead of fundamental information. HFT will have no impact on long-term price discovery. All these analysis indicate that HFT has some good impacts to the market in micro views, and no impaction in macro views.

(d) At least to this book, there is the fourth impact: the **academic** one. In the 1960s and 1970s, technical analysis was widely dismissed by academics. And it is still considered by many academics to be a pseudoscience. We use stationary process to study technical analysis since eight years ago. We discovered already that [14, 16, 19, 20, 31–33] some technical indicators can be justified by statistics of stationary processes if the popular modern financial mathematical models are valid. However, we could not find a way to show that technical analysis may lead to statistical arbitrages until we work on the high-frequency data. As we have already shown in the previous chapters, the technical analysis is indeed a useful tool in HFT. Unfortunately, the high-frequency data are quite often too expensive to academic research.

There were some system errors in HFT trading, which cause some sudden incorrect price in the market. That is one of debates on HFT. However, one should not throw away the apple because of the core. If we look into the history, there were more mistakes made by manual operation than that of HFT. And if the system of some HFT funds had made mistakes, they would be soon removed from the market, and the majorities of the HFT systems are correct and assist market to recover from such kinds of mistakes quickly. Therefore, the HFT system error is not a big issue and can be ruled out by perfecting the regulation. That is one of the major purposes for us to publish this book. Program trading is mainly performed by firms, in which the participants are bounded by their confidentiality agreements. Therefore, many strategies and algorithms of HFT are not known by academics. However, without a thorough study of HFT, how we can perfect the regulation?

It is also in debating that HFT made a lot of profit without any contribution to the real society and economy. It looks that HFT takes free lunch from the market. Actually, a HFT firm pays much more for technical improvement and transaction fees than a traditional trading firm. We may remember similar debates on bank and capital market hundreds of years ago. And after crisis of 2008, such kinds of debates returned. Greedy banker and fund managers become a symbol of evilness. However, it is obvious that the whole economy is relying on the financial service, and HFT has been providing liquidity to the market, and lower down the cost. Does anybody want to return to 10 years ago, to pay 10 times commission to exchanges?

The high profit of HFT has attracted a lot of wise brains, and a final debate is on losing these brains in the other useful fields. That is somewhat true by now, although the number of researchers in HFT is only a small fraction of financial mathematicians. Things may change in the future. As more and more participants come to HFT, and more and more competitions, the average profit rate of HFT will continuously come down to some average rate finally. That is the way the market has to become more efficient. And everyone will calm down and hunt for another opportunity in the future.

We can see that any holding back to the progress of HFT is futile. HFT is a consequence of development of financial market and IT technology. HFT is based on the random fluctuation of the market, which is just like the fluctuation of the ocean surface. One can make profit from the fluctuation of the wave (to generate the electricity, to surf and so on.) but one cannot stop the fluctuation. HFT is a highly competitive area, of which any advancement can lead to a very different profit. Just like a war may push technology advancing dramatically, HFT is a peaceful war with similar effect in the financial market. A lot of hardware, software and communication technologies have been invented, which may be used outside of HFT in the near future. And HFT makes an important application of mathematics, artificial intelligence and related fields. We can say that HFT is a science which is playing an important role in the history of financial market and with possible impacts in the other places.

References

1. Will Acworth (2013). "FIA Annual Volume Survey: Trading Falls 15.3% in 2012". Available at http://www.futuresindustry.org/downloads/FI-2012_Volume_Survey.pdf.
2. Irene Aldridge (2010). "What is High Frequency Trading, After All?" *Huffington Post*. July 8, 2010 (Retrieved August 15, 2010).
3. Andersen T. G. (1996). "Return Volatility and Trading Volume: An Information Flow Interpretation of Stochastic Volatility", *The Journal of Finance*, 51(1), 169–204.
4. Si Bao, Weian Zheng and Yu Zhou (2013). "Application of Stationary Technical Indicator in High-Frequency Trading Based on MACD", *Journal of East China Normal University (Natural Science)*, September(5), 152–160 (in Chinese, abstract in English).
5. George David Birkhoff (1931). "Proof of the Ergodic Theorem", *Proceedings of the National Academy of Sciences USA*, 17(12), 656–660.
6. Constance Brown (2012). *Technical Analysis for the Trading Professional*, 2nd Edition, The McGraw-Hill, ISBN 978-0-07-175914-4 (USA).
7. Shi Chen, Shujin Wu and Weian Zheng (2013). "ETF Arbitrage Research on China Financial Markets", *Journal of East China Normal University (Natural Science)*, September(5), 144–151 (in Chinese, abstract in English).
8. CFFEX Official Web Site. Available at http://www.cffex.com.cn.
9. CZCE Official Web Site. Available at http://www.czce.com.cn.
10. DCE Official Web Site. Available at http://www.dce.com.cn.
11. Robert D. Edwards and John Magee (1948). Technical Analysis of Stock Trends (2011 Reprint of 1958 edition, Martino Fine Books, USA).
12. Paul R. Halmos (1956). *Lectures on Ergodic Theory*, AMS Chelsea Publication, Providence, Rhode Island, USA.

13. Harrison Hong and Jeremy C. Stein (1999). "A Unified Theory of Underreaction, Momentum Trading, and Overreaction in Asset Markets", *The Journal of Finance*, 54(6), 2143–2184.
14. Xudong Huang (2008). PhD thesis presented to East China Normal University (in Chinese, abstract in English).
15. Rob Lati (2009). "The Real Story of Trading Software Espionage", *Advanced-Trading.com*, July 10, 2009.
16. Wen Li (2006). Master thesis presented to East China Normal University (in Chinese, abstract in English).
17. Chang Liu (2013). PhD thesis presented to East China Normal University (in Chinese, abstract in English).
18. Chang Liu and Weian Zheng (2013). "Transaction Price Analysis and High-Frequency Trading", *Journal of East China Normal University (Natural Science)*, September(6), 14–21.
19. Wei Liu (2006). PhD thesis presented to East China Normal University (in Chinese, abstract in English).
20. Wei Liu, Xudong Huang and Weian Zheng (2006). "'Black–Scholes' model and Bollinger Bands", *Physica A*, 371(2), 565–571.
21. Andrew W. Lo (2010). *Hedge Funds: An Analytic Perspective*, Revised and Expanded Edition), Princeton University Press, p. 260.
22. John von Neumann (1932). "Proof of the Quasi-Ergodic Hypothesis", *Proceedings of National Academy of Sciences USA*, 18(1), 70–82.
23. Richard W. Schabacker (1930). *Stock Market Theory and Practice*, B. C. Forbes Publishing Company, New York.
24. Richard W. Schabacker (1932). *Technical Analysis and Stock Market Profits*, 2005 edition, Harriman House Ltd., Hampshire.
25. SHFE Official Web Site. Available at http://www.shfe.com.cn.
26. Shuai Wang and Weian Zheng (2012). "A Discussion of Hong-Stein Model", *Science China Mathematics*, 55(11), 2367–2378.
27. Shuai Wang (2013). PhD thesis presented to East China Normal University (in Chinese, abstract in English).
28. Zhaodong Wang *et al.* (2004). "Futures Trading Data Exchange Protocol", *Financial Industry Standard in China*, JR/T0016-2004.
29. J. Welles Wilder Jr. (1978). "New Concepts in Technical Trading Systems", *Trend Research*, June, ISBN-10: 0894590278.
30. Shujin Wu (2012). *Colloquium Talk, School of Finance and Statistics*, East China Normal University, Taicang (January 6).

31. Song Xu (2011). PhD thesis presented to East China Normal University (in Chinese, abstract in English).
32. Weian Zheng (2007). Keynote Speech, CMF'2007, Wuhu, China.
33. Wei Zhu (2006). Master thesis presented to East China Normal University (in Chinese, abstract in English).

Index

σ-field, 87–90, 95
σ-field of random events, 87–90
σ-field of random events up to time, 87

A

ADXR, 133, 134, 136
alpha arbitrage, 42
alternative hypothesis, 98
Application Programming Interface (API), 24, 63
arbitrage-free hypothesis, 160
Ask Price, 22, 23, 41, 43, 44, 48, 68, 108, 155, 157
Ask Volume, 22, 23, 43, 44
assumption with validation, 50
autocorrelation function (ACF), 103
Average Direction Movement Index (ADX), 134
Average Directional Movement Indicator, 134
Average True Range, 134

B

behavioral finance, 3, 6, 93, 159, 163, 164
Bid Price, 22, 23, 43–46, 48, 68, 108, 153, 155, 157
Bid Volume, 22, 23, 43–46
big data, 47
Black–Scholes equation, 160, 161
Black–Scholes formula, 160, 161
Bollinger Bands, 112, 120–123
Borel set, 95
bounded variation, 100
Brownian motion, 92–94, 99, 120, 127, 129, 160, 161

C

calendar spread, 40–43, 47, 53
Centralized Counterparty (CCP), 29
central limit theorem, 92, 97
China Financial Futures Exchange (CFFEX), 5, 7, 144
clear information, 64
client position limit, 33, 36
close by, 34, 36, 57
Close Price, 22, 30, 46
co-location hosting, 167
cointegration, 104
complement, 86
computational finance, 6, 159, 162, 164, 165
conditional probability, 89, 96
continuous, 11, 12, 15, 21, 23, 25, 34, 36, 57, 64, 75, 84, 88, 89, 91, 92, 95, 101, 142–144, 161
countable, 86, 87
cross-market arbitrage, 40, 42, 47, 53, 63, 75, 81
cross-product arbitrage, 40, 42, 43, 47, 53

D

Dalian Commodity Exchange (DCE), 7, 42
death cross, 120

175

deterministic arbitrage, 38
direct market access (DMA), 24, 62
direction indices, 134
Directional Movement Index (DMI), 133, 134
discrete random variable, 88, 91
discrete time stationary process, 94, 95
disjoint, 86, 87, 89
Dow theory, 111, 112
dualside market data, 22, 23

E
efficient-market hypothesis (EMH), 111
elementary event, 86–88, 92, 94, 98
elementary space, 86
ergodic, 2–4, 6, 85, 93, 95–97, 101, 110, 113, 115, 141, 143, 145, 148, 163
exponential moving average, 115, 117

F
FAK (Fill and Kill), 64
fast stochastic oscillator, 132
filtering problem, 98, 100
filtration of σ-fields, 87
financial engineering, 6, 159
financial mathematics, 159
finite state automation (FSA), 69
flexibility consideration, 76
flow control, 28, 71, 74
flow control rule, 74
FOK (Fill or Kill), 66
framing, 163
full stochastic oscillator, 132
fundamental analysis, 109, 111, 168
futures dedicated account, 33
futures trading data exchange protocol (FTD), 24

G
Gaussian random variable, 89, 90, 92, 100, 164
geometric Brownian motion, 93, 120, 127, 129, 160, 161
G-measurable, 88, 90
golden cross, 120

H
heuristics, 163, 164
high-frequency trading (HFT), 1, 7, 37, 41, 59, 85, 103, 109, 141, 155, 167
highest price, 16, 22, 74
HKEX, 7
Hypothesis of Economic Man, 30
hypothesis testing, 97

I
iceberg order, 56
illegal price detection, 70
increasing process, 92, 93, 100
independent, 28, 29, 77, 81, 90–93, 95, 99, 164
indicator of the event, 91
instrument ID, 21, 43
instrument information, 64, 83
internal response time, 2, 59–61, 70, 71, 77, 81, 82
intersection, 15, 19, 20, 86–88, 90
invertible, 94
Ito-Riemann sum, 142, 143

J
joint event, 86
Just in Time (JIT), 78

K
Kalman–Bucy filter, 100

L
lag-d difference operator, 102
last price, 6, 15, 19, 21, 22, 32, 36, 41, 43, 44, 47, 155
learning curve consideration, 76
least recently used (LRU), 84
left-continuous, 142
left-limits, 142
leg multiplier, 44–46
legs, 40, 44, 45, 48, 69
liquidity, 5, 52–54, 69, 168, 169
logarithmic return, 4, 109, 110, 112–116, 128, 135, 138, 141, 143–146
lowest price, 16, 22, 74

M

mandatory order, 67–70, 73
margin call and force close rule, 34
Mark to Market rule, 29, 30, 32
market data interface, 63–65
market data multiplier, 45, 47
market inefficiencies, 164
martingale, 98–100, 143
mathematical expectation, 88
mathematical finance, 6, 143, 159, 160, 162, 165
Maximum Trade Volume Rule, 17
measurable with respect to F, 88
member position limit, 33, 36
Minimum Remain Volume Rule, 18
mining with interpretation, 50, 51
miss, 40, 48
money control, 70, 73
moving average convergence–divergence (MACD), 112, 123, 149
MV (minimal volume), 66

N

non-multiple force close, 34
n-period moving average, 117
null event, 86
null hypothesis, 98, 116

O

open interest, 22, 23, 34, 36
open price, 15, 22, 30, 32, 46
order cancellation rate, 70
order local ID (OrderLocalID), 25
order system ID (OrderSysID), 25
order/trade interface, 63, 64
OTC, 1

P

parabolic stop and reverse (Parabolic SAR), 138
passive trading, 48, 49, 52
performance consideration, 76
Poisson process, 93, 94
position limit, 13, 33, 34, 36, 70, 72, 73
position profit, 29, 31, 32, 73
potential positions by inserted orders, 72
potential positions by inserting orders, 72
potential positions by strategy orders, 73
previous close price, 22
previous settlement price, 22
price discovery, 168
probability measure, 87, 89
probability space, 86, 87
probability-preserving mapping, 94
pseudo instrument, 43
push mode, 63, 64

Q

query mode, 63, 64

R

random events, 86–90, 99
random noise component, 102, 104
random variable, 4, 88–93, 95, 98, 100, 161, 164
rate of change (ROC), 112, 126
realization, 92, 95
relative frequency, 91, 92, 96, 110, 113, 115
Relative Strength Index (RSI), 112, 127
reliability consideration, 76
removal, 48, 49, 52, 69
right-continuous, 142–144
right-continuous step-function, 142
right-limits, 142

S

sample path, 92, 93, 95–98, 100, 101, 110
seasonal component, 102
self trading detection, 70
self trading detection rule, 74
semi-martingale, 98–100, 143
settlement price, 8, 22, 29–32
Shanghai Futures Exchange (SHFE), 7
Shanghai Futures Information Tech (SFIT), 28
Shanghai Gold Exchange (SGE), 11
Sharpe ratio, 2, 147–149
simple hedge, 143, 145, 146, 148, 149
single order limit, 70

single side market, 36
slow stochastic oscillator, 132
smoothed MACD, 151
spot-futures spread, 41
spread order, 40
standard deviation, 82, 89, 105, 121, 147, 161
stationary process, 3, 85, 93–98, 100, 101, 106, 109, 110, 113, 115, 120, 138, 139, 141, 169
statistical arbitrage, 3, 38–40, 104, 106, 108, 151, 155, 160, 169
statistical finance, 6, 159, 162, 163
stochastic integral, 142, 161
stochastic oscillator, 129, 130, 132
stochastic process, 4, 92–96, 99, 101, 110, 142
stopping time, 98
strategy analyst, 55
strategy orders, 66, 67, 69, 70, 73–75
strategy spider, 54
strong law of large number, 91, 93, 95
strongly (or strictly) stationary process, 97
sure event, 86

T
team structure consideration, 76
technical analysts, 109, 112
tentative order, 67–69
the middle price rule, 15
the strong ergodic theorem, 2–4, 6, 85, 95–97, 101, 110, 113, 115, 141, 143, 145, 148, 163
ticker tape trading, 37, 49, 50, 52, 53
time series, 4, 85, 101–104, 110, 117, 162, 163
trade volume, 15, 17, 18, 20–22, 32, 36, 68
trading interface, 24, 28, 29, 61–67, 81
trend component, 102, 104
true range, 134
turnover, 7, 22, 36

U
union, 51, 86, 87, 89, 98
uniside market data, 22, 23
update time, 21

V
variance, 89, 92, 95, 161
volatility, 138, 161, 168

W
waiting order, 67, 69, 70
weakly stationary process, 97, 98
Wiener filter, 100

Z
Zhengzhou Commodity Exchange (CZCE), 7

Made in United States
North Haven, CT
24 March 2024